AF352237

CHARACTERIZATION OF CATALYTIC MATERIALS

MATERIALS CHARACTERIZATION SERIES
Surfaces, Interfaces, Thin Films

Series Editors: C. Richard Brundle and Charles A. Evans, Jr.

Series Titles

Encyclopedia of Materials Characterization, C. Richard Brundle,
Charles A. Evans, Jr., and Shaun Wilson

Characterization of Metals and Alloys, Paul H. Holloway and
P. N. Vaidyanathan

Characterization of Ceramics, Ronald E. Loehman

Characterization of Polymers, Ho-Ming Tong, Steven P. Kowalczyk,
Ravi Saraf, and Ned J. Chou

Characterization in Silicon Processing, Yale Strausser

Characterization in Compound Semiconductor Processing, Yale Strausser

Characterization of Integrated Circuit Packaging Materials,
Thomas M. Moore and Robert G. McKenna

Characterization of Catalytic Materials, Israel E. Wachs

Characterization of Composite Materials, Hatsuo Ishida

Characterization of Optical Materials, Gregory J. Exarhos

Characterization of Tribological Materials, William A. Glaeser

Characterization of Organic Thin Films, Abraham Ulman

CHARACTERIZATION OF CATALYTIC MATERIALS

EDITOR

Israel E. Wachs

SERIES EDITORS

C. Richard Brundle and Charles A. Evans, Jr.

MOMENTUM PRESS, LLC, NEW YORK

Characterization of Catalytic Materials
Copyright © Momentum Press, LLC, 2010

All rights reserved. No part of this publication may be reproduced, stored in a retrieval system, or transmitted in any form or by any means—electronic, mechanical, photocopy, recording or any other except for brief quotations, not to exceed 400 words, without the prior permission of the publisher

First published by Butterworth-Heinemann in 1993
Copyright ©1992, by Butterworth-Heinemann, a division of Reed-Elsevier, Inc.

Reissued volume published in 2010 by
Momentum Press®, LLC
222 East 46th Street, New York, N.Y. 10017
www.momentumpress.net

ISBN-13: 978-1-60650-184-9   (hard back, case bound)
ISBN-10: 1-60650-184-4   (hard back, case bound)

ISBN-13: 978-1-60650-186-3   (e-book)
ISBN-10: 1-60650-186-0   (e-book)

DOI forthcoming

Interior Design by Scribe, Inc.

10 9 8 7 6 5 4 3 2 1

Printed in Taiwan R.O.C.

Contents

BULK METAL OXIDES

SUPPORTED METAL OXIDES

ALUMINA PILLARED CLAYS: METHODS OF PREPARATION AND CHARACTERIZATION

APPENDIXES: TECHNIQUES SUMMARIES

Preface to the Reissue of the Materials Characterization Series

The 11 volumes in the Materials Characterization Series were originally published between 1993 and 1996. They were intended to be complemented by the *Encyclopedia of Materials Characterization*, which provided a description of the analytical techniques most widely referred to in the individual volumes of the series. The individual materials characterization volumes are no longer in print, so we are reissuing them under this new imprint.

The idea of approaching materials characterization from the material user's perspective rather than the analytical expert's perspective still has great value, and though there have been advances in the materials discussed in each volume, the basic issues involved in their characterization have remained largely the same. The intent with this reissue is, first, to make the original information available once more, and then to gradually update each volume, releasing the changes as they occur by on-line subscription.

C. R. Brundle and C. A. Evans, October 2009

Preface to Series

This Materials Characterization Series attempts to address the needs of the practical materials user, with an emphasis on the newer areas of surface, interface, and thin film microcharacterization. The Series is composed of the leading volume, *Encyclopedia of Materials Characterization*, and a set of about 10 subsequent volumes concentrating on characterization of individual materials classes.

In the *Encyclopedia*, 50 brief articles (each 10 to 18 pages in length) are presented in a standard format designed for ease of reader access, with straightforward technique descriptions and examples of their practical use. In addition to the articles, there are one-page summaries for every technique, introductory summaries to groupings of related techniques, a complete glossary of acronyms, and a tabular comparison of the major features of all 50 techniques.

The 10 volumes in the Series on characterization of particular materials classes include volumes on silicon processing, metals and alloys, catalytic materials, integrated circuit packaging, etc. Characterization is approached from the materials user's point of view. Thus, in general, the format is based on properties, processing steps, materials classification, etc., rather than on a technique. The emphasis of all volumes is on surfaces, interfaces, and thin films, but the emphasis varies depending on the relative importance of these areas for the materials class concerned. Appendixes in each volume reproduce the relevant one-page summaries from the *Encyclopedia* and provide longer summaries for any techniques referred to that are not covered in the *Encyclopedia*.

The concept for the Series came from discussion with Marjan Bace of Manning Publications Company. A gap exists between the way materials characterization is often presented and the needs of a large segment of the audience—the materials user, process engineer, manager, or student. In our experience, when, at the end of talks or courses on analytical techniques, a question is asked on how a particular material (or processing) characterization problem can be addressed the answer often is that the speaker is "an expert on the technique, not the materials aspects, and does not have experience with that particular situation." This Series is an attempt to bridge this gap by approaching characterization problems from the side of the materials user rather than from that of the analytical techniques expert.

We would like to thank Marjan Bace for putting forward the original concept, Shaun Wilson of Charles Evans and Associates and Yale Strausser of Surface Science Laboratories for help in further defining the Series, and the Editors of all the individual volumes for their efforts to produce practical, materials user based volumes.

C. R. Brundle   C. A. Evans, Jr.

Preface to the Reissue of
Characterization of Catalytic Material

This comprehensive volume on catalytic materials, catalytic properties, and the techniques needed to characterize both materials and properties over the wide range involved, was put together by 11 individual experts, split between academia, oil company research and engineering laboratories, and a national laboratory. Though, of course, there have been technological advances in the areas covered since the original publication, the methodology discussed for characterization and evaluation of catalysts remains as valid today as it was then. After reissuing the volume in close to its original form, it is our intent to release updates to individual chapters, plus new material, as on-line downloads, as they become available.

C. R. Brundle and C. A. Evans, Jr., January 2010

Preface

Heterogeneous catalysis has undergone a revolutionary change in the past two decades due to the development of sophisticated characterization methods that provide fundamental information about the catalyst bulk structures, surfaces, and their properties. For the first time, these characterization methods have allowed researchers to "see" the surfaces of catalytic materials, their bulk structures (crystalline as well as amorphous phases), the influence of the process conditions on the catalytic material, as well as the effect of different synthesis methods. This new information has tremendously advanced our understanding of catalytic materials and their properties. These characterization methods have become our "eyes" and are indispensible in the development of new catalytic materials. It is hard to conceive of a modern heterogeneous catalysis activity, be it research or manufacturing, without the aid of these new characterization techniques.

Catalytic materials exist in various forms (bulk metals and alloys, supported metals, bulk metal oxides, supported metal oxides, bulk metal sulfides, supported metal sulfides, zeolites, molecular sieves, and pillared clays) which, consequently, require somewhat different characterization approaches. Thus, catalytic scientists and engineers specializing in one area of heterogeneous catalysis may not be intimately familiar with other areas of heterogeneous catalysis since vastly different characterization methods may be required. For example, catalytic scientists and engineers who typically were involved with supported metal catalysts find that they have to familiarize themselves with rather different characterization methods when they wish to study metal oxide catalytic materials (bulk metal oxides, supported metal oxides, zeolites, molecular sieves, and pillared clays). Similarly, scientists and engineers new to the area of heterogeneous catalysis need some guidance as to the applicable characterization methods. To satisfy these requirements this volume is organized by type of heterogeneous catalytic material, and emphasizes the different properties that can be determined by various characterization methods for each class of material. The Series, of which this volume is part, has the title "Materials Characterization: Surfaces, Interfaces, Thin Films" and the general concept, or intent is to provide a practical guide to people working in various materials classes within the framework of this title. In the case of *Characterization of Catalytic Materials*, the emphasis is not always on surfaces, thin films, or interfaces. For those catalytic materials that possess the active component at the surface, the surface characterization methods are critical, even though we currently have still some way to go before we can say that all the important information can be readily accessed. For those materials that possess the active catalytic component in the bulk, the bulk characterization methods are critical. An attempt has been made to focus many of the chapters on the catalytic properties to be measured rather than on the details of the characterization techniques. Thus, the information required in order to understand the structure and performance of a given catalytic material is emphasized. For ease of reference, short summaries of the techniques discussed in the book are presented in a collection of appendixes. This book is a practical guide for the characterization of catalytic materials as it is done today.

I want to thank the authors who made this book possible and were responsible for the individual chapters on different catalytic materials. It was a pleasure, as well as an education, to work with individuals who are experts in their respective fields. The high quality of the chapters greatly simplified my task as Editor. I also wish to acknowledge our Managing Editor, Lee Fitzpatrick, who orchestrated the publication in an efficient and professional manner.

Israel E. Wachs

Contributors

James F. Brazdil
BP Research International
Cleveland, OH

Bulk Metal Oxides

Jean-Rémi Butruille
Michigan State University
East Lansing, MI

Alumina Pillared Clays:
Methods of Preparation and Characterization

Michel Daage
Exxon Research & Engineering Co.
Annandale, NJ

Bulk Metal Sulfides

Mark E. Davis
California Institute of Technology
Pasadena, CA

Zeolites and Molecular Sieves

John B. Higgins
Mobil Central Research Laboratory
Princeton, NJ

Zeolites and Molecular Sieves

George Meitzner
Exxon Research & Engineering Co.
Annandale, NJ

Supported Metals

Thomas J. Pinnavaia
Michigan State University
East Lansing, MI

Alumina Pillared Clays:
Methods of Preparation and Characterization

Roel Prins
Swiss Federal Institute of Technology
Zürich

Supported Metal Sulfides

Johannes Schwank
University of Michigan
Ann Arbor, MI

Bulk Metals and Alloys

Kohichi Segawa
Sophia University
Tokyo

Supported Metal Oxides

Israel E. Wachs
Lehigh University
Bethlehem, PA

Supported Metal Oxides

Bulk Metals and Alloys

JOHANNES SCHWANK

Contents

1.1 Introduction

The Role of Metals and Alloys in Catalysis

Metals and alloys play a key role in catalytic technology, especially in reactions involving hydrogen transfer and in hydrocarbon conversion reactions. This chapter discusses the characterization of catalytic metals and alloys in their bulk form. Many catalytic applications require catalysts with high surface area to provide adequate contact between the reactants and the catalyst. Therefore, a large number of catalytic processes rely on supported-metal catalysts, in which the metal particles are well dispersed on a high surface area support material. However, unsupported bulk metals or alloys are sometimes employed in the form of films, foils, wires, or powders. Because of their high catalytic activity, platinum group metals find the widest application. The reactions of hydrocarbons on Pt surfaces include hydrogenation and dehydrogenation; hydrogenolysis of C–C, C–S, and C–N bonds; isomerization; and cyclization reactions.

On an industrial scale, ammonia is oxidized to nitric acid in the presence of a Pt–Rh wire gauze or Pd–Au alloy wires. Palladium–gold alloys are very effective catalysts for the selective hydrogenation of unsaturated hydrocarbons, such as the hydrogenation of acetylene to ethylene. Catalytic hydrogenation can also be accomplished over Raney nickel or cobalt catalysts. Silver gauze or granular silver screened to a specific particle size can be used to convert methanol to formaldehyde. In the Fischer–Tropsch process at Sasol, fused-iron catalysts are used for the hydrogenation of carbon monoxide. Under reaction conditions, the iron catalyst is converted to iron carbide, and surface carbon deposition occurs. Nickel, cobalt, and ruthenium are also active catalysts for CO hydrogenation. Iridium-promoted platinum gauze or rhodium finds application in the Andrussow process, in which hydrogen cyanide is manufactured from ammonia, methane, and oxygen. The main component of ammonia synthesis catalysts is iron. Pure iron is an effective catalyst, but it becomes rapidly deactivated unless small amounts of promoter oxides are present.

Preparing bulk metal catalysts in the form of metal foils, wires, and gauzes is fairly straightforward. For fundamental research purposes, single-crystal model catalysts are widely used. These can be prepared from the melt, from vapor, from supersaturated solutions, via electrodeposition, by thermal diffusion, and by other methods.[1]

A second metal component may be added to a metal catalyst to systematically modify the size and, in some instances, the electronic structure of catalytic surface sites.[2] Bimetallic systems can be prepared which are not necessarily in thermodynamic equilibrium, exhibiting structures and compositions deviating from bulk phase diagrams. Depending on the preparative conditions, it is even possible to arrive at amorphous structures. Structural defects can greatly affect the catalytic properties of metals and bimetallic systems. Numerous studies in the catalytic and surface science literature explore the influence of a second metal component on the bulk and surface structure, the adsorption characteristics, the surface coverage with reactive intermediates, and consequently, the activity and selectivity for catalytic reactions. Figure 1.1 shows the effect of systematically covering a Pt(111) crystal surface with gold.[3] Note that bimetallic catalyst surfaces can have very different product selectivities compared with monometallic surfaces.

The second metal component can also alter the reducibility of the catalyst and the deactivation behavior. For example, bimetallic catalyst systems in many hydrocarbon conversion reactions have shown superior activity maintenance characteristics while giving desirable product selectivity. Aside from their industrial applications, bulk bimetallic catalysts or alloys play a major role in fundamental research, where one can study the effect of the second metal component undisturbed by contributions from the support. One cannot always assume that the surface of an alloy has the same elemental composition as the bulk. In ideal bimetallic or multimetallic systems, it is possible to theoretically predict the thermodynamic equilibrium composition of the surface as a function of the bulk composition and temperature. In general, the metal with the lower energy of sublimation tends to segregate at the surface.

1.2 Preparation of Bulk Alloy or Bimetallic Catalysts

Unsupported alloy or multimetallic catalysts can be prepared by several fairly simple procedures. For example, to prepare alloys in powder form, mixtures of metal salts

BULK METALS AND ALLOYS   Chapter 1

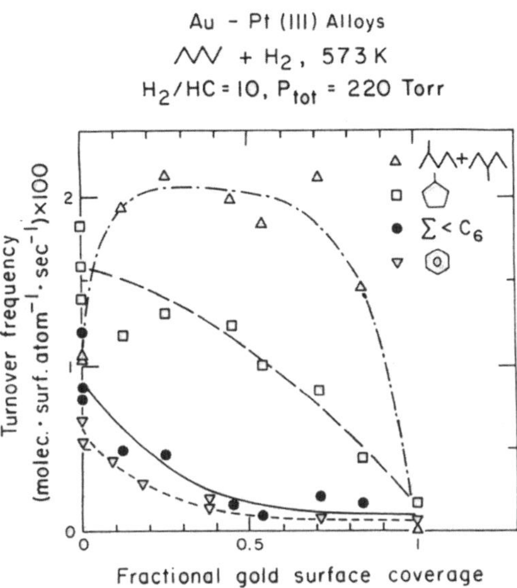

Figure 1.1   Rate of product formation from *n*-hexane over Pt(111) surfaces progressively diluted with gold. (Reproduced with permission from Reference 3.)

such as metal chlorides are reduced in flowing hydrogen gas or in hydrazine solutions[4] or some other suitable reducing medium. To ensure that the resulting alloy is homogeneous, one evaporates the solvent until a dry halide salt mixture is obtained, keeping the reduction temperature as low as possible to prevent sintering and keeping the surface area of the powder as large as possible. Nonnoble metal alloy catalysts are usually made from their corresponding carbonates, nitrates, or hydroxides, which are first calcined to convert them into an oxide mixture and then reduced in flowing hydrogen to generate an alloy. Reduction of the precursor salt mixtures can also be carried out in aqueous or nonaqueous solutions, with sodium borohydride, hydrazine, formaldehyde, or hypophosphorous acid as reducing agents. One often finds traces of sodium and boron impurities in these types of catalysts. Like monometallic catalysts, alloy catalysts can be prepared by the evaporation of suitable metals. This preparation technique is frequently used to make alloy films of high purity.

Metal catalysts, such as Raney nickel or cobalt catalysts, are widely used for catalytic hydrogenation reactions. These Raney-type catalysts are prepared from Ni–Al or Co–Al alloys; they are activated by leaching the aluminum out of the alloy in a sodium hydroxide solution at a pH of approximately 14.[5] The leaching process leaves behind a porous skeleton of Ni or Co, with a greater surface area than the original alloy. Amorphous alloys, such as Fe–Ni,[6] AuZr,[7] or FeB,[8] can be prepared by methods including vapor deposition, sputtering, electroplating, chemical plating, and rapid quenching of melts. These amorphous alloys represent nonequilibrium systems, and their use as catalysts is restricted to temperatures below their crystallization temperature. Once these alloys are exposed to temperatures above their crystallization temperature, their amorphous character may be lost and their catalytic activity may be drastically altered. An indication of the absence of long range crystalline order in amorphous metal films is their much higher electrical resistivity compared with crystalline metals.[9]

A battery of physical and physicochemical characterization techniques is employed to discover the key factors contributing to the catalytic behavior of a given metal or alloy surface. Most of these methods may apply to supported catalysts. In this chapter, methods are highlighted that are of special importance for the characterization of bulk metal systems. Aside from the standard spectroscopic methods, the measurements of electrical conductivity and of the magnetic properties of bulk metal catalysts can provide very useful information. These methods are especially useful in the case of alloy catalysts. Figure 1.2 shows the magnetization of a copper–nickel alloy changing as a function of composition.

1.3   Bulk Metal Characterization Methods

Bulk Chemical Analysis

For catalytic applications it is important to have reliable information about the chemical composition of the catalytic material, and special attention must be paid to impurities which may have been present in the precursor compounds from which the catalyst was manufactured. Additional impurities may be introduced during pretreatment of the catalytic material, for example, from containers and reaction vessels used during pretreatment, calcining, and reduction. Atomic

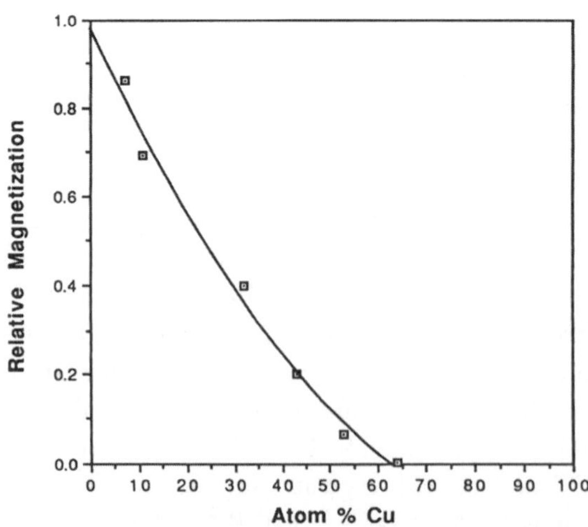

Figure 1.2   Magnetization of copper–nickel alloys as a function of composition. (Adapted with permission from Reference 11.)

BULK METALS AND ALLOYS   Chapter 1

absorption spectrochemical analysis and atomic emission–inductively coupled plasma methods are mainly used in bulk chemical analysis. X-ray fluorescence and neutron activation analysis are also widely used methods that have the advantage of not requiring the dissolution of the metal or of the alloy. X-ray fluorescence analysis is most sensitive for elements with high atomic weight.

Bulk chemical information can also be obtained using electron probe microanalysis (EMPA). EMPA instruments generate a high X-ray yield through electron bombardment of the sample and are typically equipped with an optical microscope for viewing the sample. One of the most attractive applications of EMPA is the measurement of elemental composition, including impurities, over a cross-section of a catalyst specimen.

Determination of Crystal Structure

X-ray diffraction is a powerful method used in the determination of the bulk crystallographic structure of catalytic materials; it is also widely used for the characterization of bulk metals and alloys.[10] From X-ray diffraction patterns, lattice spacings for the various crystallographic planes in a metal or alloy can be determined with great accuracy. One can determine all metallic or intermetallic phases present in a bulk metal sample in a straightforward manner by comparing the unknown X-ray diffraction pattern with known patterns of metals and alloys. In substitutional alloys that form a continuous series of solid solutions, the lattice spacings vary with composition, as demonstrated in Figure 1.3 for a series of copper–nickel alloys.[11]

X-ray diffraction lines are generally very narrow for bulk samples in the form of single crystals, thin films, or powders with large grains, with the actual width of a

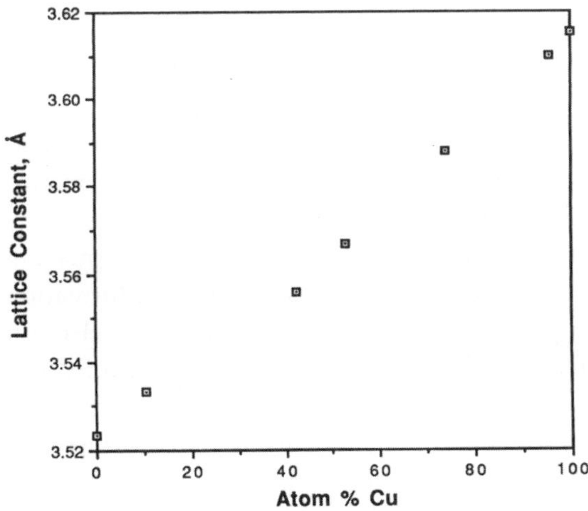

Figure 1.3 **Lattice constants of copper–nickel alloys as a function of composition. (Adapted with permission from Reference 11.)**

diffraction line dependent on instrument limitations. Polycrystalline materials with average crystallite sizes of less than 100 nm produce diffraction line broadening because of incomplete destructive interference of the X-rays. From the extent of the line broadening, one can estimate the average crystallite size, L_{hkl}, by means of the Scherrer equation:

$$L_{hkl} = \frac{K\lambda}{\beta_{hkl} \cos\theta}$$

where K is a constant that depends on the crystal geometry and is often set equal to 1. The parameter β represents the line broadening after correcting for instrumental contributions to a finite line width. Various procedures exist for separating the instrumental contribution from the broadening due to the particle size, including a method based on a Fourier transform.[12, 13] In bulk samples, lattice strain can contribute some additional broadening of the lines.

Morphology and Microstructure

Modern analytical electron microscopes offer a variety of methods for determining the morphology and microstructure of bulk metal catalysts. The most widely used electron beam techniques are imaging, diffraction, energy dispersive X-ray analysis (EDX), and electron energy loss spectroscopy (EELS). Imaging is possible in transmission, scanning transmission, scanning secondary electron, and scanning backscattered electron modes. The topography of a sample can be imaged by scanning electron microscopy (SEM), in which a striking three-dimensional image of the surface is produced when the number of secondary electrons which are emitted and reach the detector increases as a sample region is tilted towards the detector. The SEM method is very useful in assessing the influence of sample preparation on the quality of the surface, for example, surface flatness over extended regions as a function of annealing time; SEM can also be used to study the equilibrium forms of small metal crystallites.[14]

Conventional scanning electron microscopes require a high vacuum for operation, but the recent development of "environmental" scanning electron microscopes has provided new opportunities to image samples while they are being heated in reactive gas atmospheres at pressures up to about 10 torr.[15] The imaging mode can be coupled with energy-dispersive X-ray analysis to obtain information about the chemical nature of the sample. Metals that are irradiated by the electron beam in the microscope emit characteristic X-rays, which give information about the spatial distribution and concentration of elements in the sample. X-ray maps can be generated showing the location of the different elements in the sample. In a homogeneous alloy sample, the X-ray maps for two elements should overlap, though discrepancies in the X-ray map areas for the two elements can indicate non-uniform distribution of the alloy components within the specimen. In this regard, analytical electron microscopy has excellent spatial resolution and provides microstructural and phase information for small sample regions. This localized information complements X-ray diffraction results concerning the metallic phases present in the entire sample. Another very promising method for characterizing a sample's topography is scanning tunneling microscopy (STM), which is currently under intense development and has the potential to provide nearly atomic level resolution. Figure 1.4 is an STM picture obtained on a smooth, reconstructed Au(100) surface. The rough features represent carbon islands on the surface of gold.[16]

Transmission electron microscopy (TEM) and scanning transmission electron microscopy (STEM) in the low-resolution mode can be used to determine crystal morphology and the crystal habits of metal particles and to assess particle agglomeration and sintering. Electron diffraction studies may be performed with either standard selected-area techniques or by means of highly convergent electron probes. A television camera can be connected to the microscope to directly capture electron diffraction patterns on computer. These patterns can then immediately be compared to an on-line computer library for identification of structures.

In the high-resolution mode of TEM, structural images can be obtained with atomic resolution, particularly in high voltage instruments. The interpretation of these images is not always straightforward, since the image is a two-dimensional representation of a three-dimensional structure. Computer simulations and image

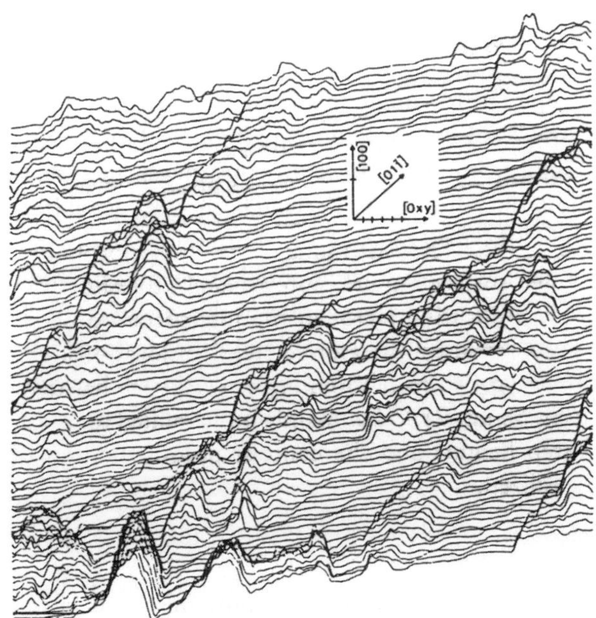

Figure 1.4   STM picture showing sharply structured carbon islands on the smooth, reconstructed Au(100) surface. The divisions on crystal axes are 5 Å apart. (Reproduced with permission from Reference 16.)

calculations, such as the multislice routine, are often used to aid in the interpretation of structure images. Lattice fringe images can be recorded on photographic film or in digital form in a computer. High resolution video recorders can also be connected to the microscope for dynamic observations. From the intensity variations in the negatives of the lattice fringe images an optical diffraction pattern can be obtained that allows one to accurately determine lattice spacings and directions. The optical diffraction pattern is essentially the Fourier transform of the lattice image; it is obtained by passing a parallel monochromatic laser beam through the image negative.

Specimen preparation for transmission electron microscopy of metals is quite involved, as a TEM specimen must be very thin and have the right size and shape to fit the microscope sample stage. An excellent review of metal specimen preparation techniques for TEM has been written by Goodhew.[17] Typically, the preparation starts with a metal disc about 3 mm in diameter, having a thickness of 100–500 μm. These discs are further thinned by methods such as ion milling or electropolishing. One of the greatest challenges is to prepare a thin enough specimen without seriously altering the structure and composition of the sample and without introducing unacceptable artifacts. Furthermore, the minuscule amount of sample that may be analyzed in the microscope makes the selection of a truly representative sample of the catalyst nontrivial. The specimen preparation technique has to be selected on the basis of the amount of specimen damage one can tolerate and the information one is trying to extract from the sample. It may not always be possible to obtain all the information needed from one specimen.

The reflection electron microscopy (REM) mode in commercial electron microscopes can be used to study the surfaces of bulk specimens in high resolution. With REM, one may distinguish the various features and structures present on the surface. The technique produces images from the reflection of electron beams diffracted from planes parallel to the surface. The REM images show the morphology and defect structures of the specimen surface but are, unfortunately, foreshortened in the beam direction. This is a complicating factor in the interpretation of the micrographs. In order to obtain information on atomic steps, dislocations, and other features from a REM micrograph, the foreshortening of the image has to be countered via a geometrical analysis. So far, REM has mainly been applied to simple cubic structures. The expected step heights and the contrast produced by more complex structures are yet to be realized.

Quantification of Surface Area

The total surface area of bulk metal catalysts is most commonly measured by the Brunauer–Emmet–Teller (BET) method,[18] for which commercial instruments are available. The BET method is based on physical adsorption of gases such as nitrogen or krypton. Bulk metal catalysts tend to have fairly low surface areas, except perhaps Raney-type powders. The pore volume and pore size distribution of the

     BULK METALS AND ALLOYS   Chapter 1

latter can be determined by mercury intrusion porosimetry.[19] A detailed discussion of pore structure determination in powders can be found in the monograph by Parfitt and Singh.[20]

Metal surface areas can also be measured by selective chemisorption of gases, where a monolayer of adsorbate is formed. Typical gases used for chemisorption on metals are H_2, O_2, CO, C_2H_2, and C_2H_4. Static volumetric methods are very time consuming, as they require the collection of many data points to determine a reliable adsorption isotherm. Flow and pulse methods are much faster and agree well with static methods if the adsorption process is very rapid and irreversible at the adsorption temperature.[21] One has to make an assumption regarding the chemisorption stoichiometry in order to relate the number of gas molecules adsorbed in the monolayer with the number of surface metal atoms. For example, it is generally assumed that one Pt surface atom chemisorbs one atom of hydrogen; however, for polycrystalline samples one has to make further assumptions regarding the types and relative contribution of crystallographic planes exposed on the surface. Then one can calculate an average site density. So, if we assume an equal mixture of Pt(100) and Pt(110) surfaces, we obtain a site density of 1.12×10^{15} sites/cm^2, corresponding to an average surface area of 0.89 nm^2 per Pt surface atom.

Surface Composition

Many techniques are available for the surface characterization of bulk metals and only a brief overview will be given here. Metal surfaces can be probed by either a particle beam or by electromagnetic or thermal energy. The output signal can consist in the emission of particles or photons of discrete energy, which can then be analyzed by suitable detectors. Many surface characterization techniques require an ultrahigh vacuum (UHV); however, it is very difficult to maintain clean metal surfaces during analysis. Even in a vacuum of 10^{-9} torr there are enough residual gas molecules in the analysis chamber of a typical spectrometer that it will take only about 15 minutes for the sample surface to be covered with a monolayer of adsorbate, if one assumes that every collision of a gas molecule with the surface results in adsorption.

A key characteristic of a bulk metal catalyst is the relationship between its bulk composition and its surface composition. In ideal bimetallic systems it is possible to derive thermodynamically the equilibrium surface composition as a function of the bulk composition and temperature. The heat effects of surface segregation can be determined from the "broken bond" model, in which it is assumed that each metal atom in the bulk forms a certain number of bonds with nearest neighbor atoms: 8 for bcc metals and 12 for fcc metals. Metal atoms on the surface have fewer nearest-neighbors, and several bonds are "broken," depending on the crystal face exposed. Metals with lower sublimation energy and, consequently, weaker metal–metal bonds tend to segregate in the surface, where less energy is used by not forming the weaker bonds. Segregation is also more prevalent on rougher surfaces with higher index planes, since there is a greater number of bonds not formed on

such surfaces. Many practical bimetallic systems are much more complicated, however, and show deviations from ideal behavior. In non-ideal systems the surface segregation may extend to more than one surface layer, and differences in atom size need
to be taken into account.[22]

The experimental determination of surface composition in non-ideal systems
is problematic. X-ray photoelectron spectroscopy (XPS) and Auger electron spectroscopy (AES) can give information about the composition and electron structure
of the first few atomic layers on the surface, but it is not a simple task to assess
the contributions of deeper layers to the XPS or Auger signals obtained from the
surface. If one assumes that only the first layer is different from the bulk, one can
use the following equation[22] to relate Auger signal intensities to the surface and
bulk concentrations:

$$\frac{I(A)}{I(B)} = R* \frac{\rho^B}{\rho^A} \frac{N_1^A x_1 + (1 - N_1^A)x_{\text{bulk}}}{N_1^B(1 - x_1) + (1 - N_1^B)(1 - x_{\text{bulk}})}$$

where $I(A)$ and $I(B)$ are Auger signal intensities for elements A and B in the alloy,
respectively, divided by the Auger signal intensity of the pure metals A and B; $R*$ is
a correction factor for back scattering; ρ^A and ρ^B represent the planar densities of
elements A and B, respectively; N_1 is the fraction of the total signal emanating from
the first layer; x_1 is the molar ratio of A in the top layer; and x_{bulk} is the molar ratio
on the bulk of the alloy. Figure 1.5 compares experimental Auger spectroscopy data
for the surface composition of Pt–Cu alloys with predicted surface compositions for
ideal alloys. Note the significant deviations from ideal behavior.

Samples for XPS and Auger electron spectroscopy should have flat surfaces; thin
metal foils are ideal. Powder samples must be pressed into the shape of a thin wafer.
To study a clean, reduced metal surface, it is necessary to transfer a sample of the
metal from a pretreatment chamber operated at higher pressures into the analysis
chamber, which runs at UHV, without exposing the sample to ambient air. Commercial transfer devices with appropriate interlock and pump systems are available.

In XPS, the effective depth of the analyzed layer is about 2–4 nm, and for Auger
electron spectroscopy, about 1–3 nm. It is possible to sputter the surface with high
energy argon ions that remove atoms from the surface, so that depth profiles can
be acquired. Depth profiling can give valuable information about the location of
impurities and is a very useful method for studying compositional changes between
the surface and the subsurface. The spatial resolution of XPS is not very good, the
signal arising from an area about 10 000 nm across. Much better spatial resolution
can be achieved in a scanning Auger microprobe (SAM), where the signal emanates
from an area as small as 20 nm across. In samples with a non-uniform distribution
of metal components, XPS tends to give an average surface composition, whereas
the SAM pinpoints the local variations in surface composition. The SAM can also
reveal the oxidation state of the surface, which can be very important because many
metals have a tendency to undergo surface oxidation when exposed to ambient air.

                                    BULK METALS AND ALLOYS   Chapter 1

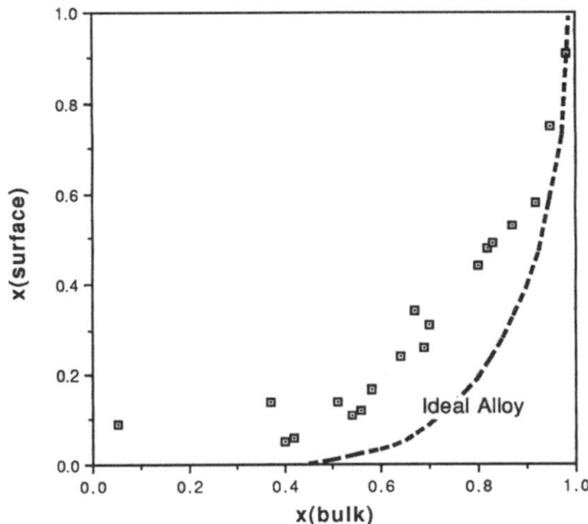

Figure 1.5   **Auger data showing the surface Pt content of Pt–Cu alloys as a function of the bulk content. The dotted line indicates the theoretically-expected surface composition of ideal Pt–Cu alloys. (Adapted with permission from Reference 22.)**

To understand the catalytic behavior of a material, it is critical for one to check for trace impurities on the surface of the material; these may not show up in bulk analysis due to their low overall concentration. Given the relatively low surface area of typical bulk metal catalysts, even a minute amount of surface impurities can have drastic effects on the catalytic behavior.

Ultraviolet photoelectron spectroscopy (UPS) provides information about the electron structure of the surface and also probes the nature and geometry of adsorbate bonds. Secondary-ion mass spectrometry (SIMS) is a highly sensitive, semiquantitative tool for surface characterization of metals that can probe the local environment of atoms on the surface by performing depth profiling. A drawback of SIMS is the possibility of changing the composition and chemical state of the surface as a consequence of the intense ion bombardment. A topographic image based on secondary ion emission may be obtained using SIMS. Often SIMS is used in conjunction with other surface analysis techniques. Similarly, ion scattering spectroscopy (ISS) can be used very effectively along with other spectroscopic characterization methods to study the surface composition of metals and alloys and to obtain structural information, especially in the case of single-crystal surfaces.

Gas–Surface Interactions

Temperature-programmed methods can provide a wealth of useful information about bulk metal systems and their interactions with gas molecules. In a typical temperature-programmed desorption (TPD) experiment, a gas is adsorbed on the surface, and then the temperature of the surface is increased as a function of time.

For single crystals, thin films, and metal foils, the temperature of the sample is raised at a typical rate of 10 K/s, and the desorption spectrum is recorded in a UHV system by a mass spectrometer. This method is referred to in surface science literature as flash desorption spectroscopy.[23] Figure 1.6 gives a good example of how TPD can be used to study the effects of alloying.[24] The position of desorption peaks is an indication of the strength of adsorption, and the number of desorption peaks can give information about the number of energetically distinct surface sites. As shown in the figure, CO is adsorbed only weakly on pure Cu, and the desorption peak occurs at low temperatures in the range of 200–225 K. On the other hand, CO is much more strongly held on pure Ni, and the desorption of CO requires much higher temperatures. When increasing amounts of Ni are added to Cu, additional desorption peaks occur besides the TPD peaks characteristic for desorption from Cu and Ni sites, indicating the presence of energetically distinct surface sites where the surface ensembles of Cu are modified by Ni atoms.

For powder samples, the experiments are often carried out in flow systems, where the desorbing species are desorbed into a stream of carrier gas instead of a vacuum and are analyzed by means of a gas chromatograph detector or a quadrupole mass spectrometer.[25] One has to be careful to take into account possible complications due to heat and mass transfer limitations or readsorption of gas. For single-crystal surfaces and well-defined thin films, work function and contact potential measurements can be taken.[26] In most cases, gas adsorption on a metal surface either increases or decreases the work function through charge transfer between the adsorbed molecules and the substrate. Several experimental methods[27] exist. Contact potential difference measurements with the Kelvin probe provide a powerful means of monitoring the change in work function of a metal surface during gas adsorption and temperature-programmed desorption.

Vibrational spectroscopies, including infrared spectroscopy (IR) and high-resolution electron energy loss spectroscopy (HREELS), probe the structure and bonding of adsorbed surface species. Bulk metal catalysts typically are opaque and do not permit adequate transmission of infrared radiation. On foils or flat surfaces of single crystals, specular reflection methods can be used; on rough surfaces, diffuse reflectance spectroscopy (DRS).[28] In DRS, the diffusely scattered radiation is collected by a hemispherical or elliptical reflector and focused into the IR detector. Extended X-ray absorption fine-structure spectroscopy (EXAFS) is a promising tool for determining the local structural parameters around an excited atom, such as coordination numbers and interatomic distance. Near-edge X-ray absorption fine-structure spectroscopy (NEXAFS) can provide detailed molecular information, such as bond tilt angles, for adsorbed species on single-crystal surfaces.

Surface Structure of Single Crystals and Metal Films

The main technique used in determining the surface structure of single crystals is low energy electron diffraction (LEED). The first layer of atoms on a surface may

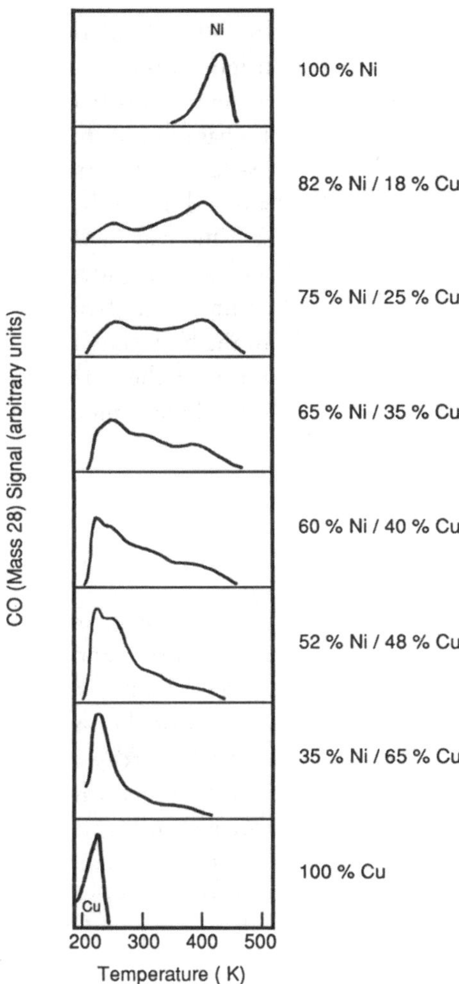

Figure 1.6 TPD spectra of CO from Ni(110) surface, Cu(110) surface, and 90% Ni–10% Cu bulk alloy with varying surface compositions indicated on the right side of each TPD spectrum. The heating rate was 8 K/s and the initial CO exposure was 1 L. (Adapted with permission from Reference 24.)

have a structure which deviates from the bulk structure, which is indicated by the appearance of extra diffraction spots in LEED patterns—diffraction spots other than those characteristic for the bulk structure. Surface structure analysis by LEED has provided strong evidence that the surface atoms in many metals and alloys are located in new equilibrium positions. In experiments with Pt single crystals, it was found that the rates of many hydrocarbon conversion reactions vary, depending on the type of crystallographic face exposed.[29] The information gained from studies on well-defined single-crystal surfaces can be used to better interpret the catalytic behavior of thin film, powder, or Raney-type catalysts. Besides identifying the surface structure of the metal, LEED can also give valuable insight into the arrangement of adsorbed overlayers with respect to the structure of the metal surface.

Vapor deposition of pure metals is ideally suited for the preparation of thin film model catalysts, which can be studied by electron emission spectroscopic methods such as XPS, ion scattering, LEED, and temperature-programmed methods. Some thin films have very special properties of catalytic interest. One can deposit an inactive metal such as gold on the surface of an active one such as ruthenium, thereby breaking up the surface ensembles of the active component. (Note that these bimetallic films contain metals which are immiscible in the bulk state.) Similarly, one can deposit an active component such as platinum on the surface of another catalytically active metal such as rhodium or palladium, thereby modifying their adsorption characteristics and catalytic activity.

1.4   Surface Composition–Structure and Catalysis Relationship

Heterogeneous catalytic reactions are controlled by the atomic-scale structure and composition of the surface and the interactions of surface sites with reactant molecules. To truly understand the activity and selectivity of a given catalytic system, one needs to have a clear idea of the structure and composition of the catalytic surface. Model studies on single-crystal surfaces and thin films give valuable insight into the structure–activity relationships governing industrial bulk metal catalysts. However, additional features influence the catalytic performance of industrial bulk metal catalysts, for instance, trace impurities and the presence of surface oxide or carbide species. Developing a detailed, atomic-scale picture of the microstructure and surface composition of bulk metal catalysts represents a challenging and promising research frontier in heterogeneous catalysis.

A better understanding of the relationships between the properties and processing of bulk metal catalysts and their microstructures is the key to the design of new catalytic materials with specific physicochemical surface properties. Advanced spectroscopic methods such as XPS, AES, UPS, and SIMS enable us to address fundamental questions with regard to surface and near-surface properties better than routine quantitative bulk analysis using atomic absorption spectroscopy, X-ray fluorescence, and neutron activation analysis. X-ray diffraction has been a powerful and accurate method for the determination of bulk crystallographic structures, but recent developments in high-resolution and analytical electron microscopy, which allow spacial resolutions approaching the atomic scale, have opened to us the realm of microstructural and chemical characterization. Especially in the case of bimetallic systems, it is very important to know how the bulk composition relates to the surface composition and how the second metal component modifies the geometry of the surface ensembles. EXAFS is a valuable tool for probing bond distances, coordination numbers, and disorder.

                                          BULK METALS AND ALLOYS   Chapter 1

Rough surfaces tend to have high index planes and possess surface site geometries which can be quite different from those encountered on smooth surfaces with low index planes; SEM and STM provide information about surface morphology, which greatly influences the catalytic behavior of bulk metal catalysts. Adsorption of probe molecules gives a reliable measure of surface area, and temperature-programmed desorption methods give insight into the energetics of adsorption sites and their modification by second metal components. Additional information about the structure and bonding of adsorbed species may be obtained through the use of IR, HREELS, and NEXAFS.

A comprehensive, multifaceted catalyst-characterization regimen provides the data necessary to better understand catalytic performance and to diagnose the reasons for loss of catalytic activity under industrial reaction conditions. Through a judicious combination of *ex* situ and in situ studies of bulk metal catalysts, one may correlate the dynamics of surface chemical and physical changes with changes in the chemical and physical environment and thereby better understand the key principles governing the catalytic behavior of bulk metal catalysts.

References

1   W. D. Lawson and S. Nielsen. *Preparation of Single Crystals*. Butterworth, London, 1958.

2   J. H. Sinfelt. *Bimetallic Catalysts—Discoveries, Concepts, and Applications*. Wiley, New York, 1983.

3   G. A. Somorjai. *Proceedings of the Robert A. Welch Foundation Conferences on Chemical Research, XXV: Heterogeneous Catalysis*. The Robert A. Welch Foundation, Houston, 1983, p. 121.

4   A. E. Newkirk and D. W. McKee. *J. Catal.* **11**, 379, 1970.

5   A. B. Stiles. *Catalyst Manufacture*. Marcel Dekker, New York and Basel, 1983, p. 131.

6   A. Yokoyama, H. Komiyama, H. Inoue, T. Matsumoto, and H. M. Kimura. *J. Catal.* **68**, 355, 1981.

7   M. Shibata, N. Kawata, T. Masumoto, and H. M. Kimura. *Chem. Lett.* 1605, 1985.

8   G. Kisfaludi, K. Lázár, Z. Schay, L. Guczi, C. Fetzer, G. Koncos, and A. Lovas. *Appl. Surface Sci.* **24**, 225, 1985.

9   S. Mader, A. S. Nowick, and H. Widmer. *Acta Met.* **15**, 203, 1967.

10   B. D. Cullity. *Elements of X-Ray Diffraction*. Addison-Wesley, Reading, Mass., 1978.

11   J. H. Sinfelt, J. L. Carter, and D. J. C. Yates. *J. Catal.* **24**, 283, 1972.

12 H. P. Klug and L. E. Alexander. *X-Ray Diffraction Procedures.* Wiley-Interscience, New York, 1974.

13 J. B. Cohen and L. H. Schwartz. *Diffraction from Materials.* Academic Press, New York, 1977.

14 J. A. Venables. *Chemistry and Physics of Solid Surfaces, IV.* (R. Vanselow and R. Howe, Eds.) Springer-Verlag, New York, Heidelberg, and Berlin, 1982, Chapter 6.

15 R. B. Bolon, C. D. Robertson, and E. Lifshin. *Microbeam Analysis—1989.* (P. E. Russell, Ed.) San Francisco Press, San Francisco, 1989, p. 449.

16 G. K. Binnig, H. Rohrer, C. Gerber, and E. Stoll. *Surface Sci.* **14**, 321, 1984.

17 P. J. Goodhew. "Specimen Preparation for Transmission Electron Microscopy of Materials," *Materials Research Society Symposium Proceedings, Vol. 115.* (J. C. Bravman, R. M. Anderson, and M. L. McDonald, Eds.) Materials Research Society, Pittsburgh, 1988, p. 51.

18 "Standard Test Method for Surface Area of Catalysts," *Annual Book of ASTM Standards.* D3663-84. American Society for Testing and Materials, Philadelphia, 1988.

19 "Standard Test Method for Determining Pore Volume Distribution of Catalysts by Mercury Intrusion Porosimetry," *Annual Book of ASTM Standards.* D4284-83. American Society for Testing and Materials, Philadelphia, 1988.

20 G. D. Parfitt and S. L. Singh. *Characterization of Powder Surfaces.* Academic Press, New York, 1976.

21 A. Hansen and H. L. Gruber. *J. Catal.* **20**, 97, 1971.

22 A. D. van Langeveld and V. Ponec. *Applic. of Surface Sci.* **16**, 405, 1983.

23 D. A. King. *Surface Sci.* **47**, 384, 1975.

24 K. Y. Yu, D. T. Ling, and W. E. Spicer. *J. Catal.* **44**, 373, 1976.

25 R. J. Cvetanovic and Y. Amenomiya. *Catal. Rev.* **6**, 21, 1972.

26 P. M. Gundry and F. C. Tompkins. *Experimental Methods in Catalytic Research.* (R. B. Anderson, Ed.) Academic Press, New York, 1968, p. 100.

27 G. Ertl and J. Küppers. *Low Energy Electrons and Surface Chemistry.* Verlag Chemie, Weinheim, 1974.

28 K. Klier. "Vibrational Spectroscopies for Adsorbed Species," *ACS Symposium Series.* (A. T. Bell and M. L. Hair, Eds.) **137**, 141, 1980.

29 G. A. Somorjai. *Proc. 8th Int. Congr. Catal. (Berlin).* **1**, 113, 1984.

Supported Metals

GEORGE MEITZNER

Contents

2.1 Introduction

There are four primary methods for characterizing metals in catalysts: chemisorption, reaction studies, X-ray diffraction, and electron microscopy. Also, X-ray absorption spectroscopy is a very powerful, increasingly-used, general method. This chapter will review these methods and the information each can provide. Other, more-specialized methods will be discussed briefly: Mössbauer spectroscopy, photoelectron spectroscopy (including X-ray and ultraviolet photoemission), and magnetic methods.

Many of the heterogeneous catalysts used in reducing environments include a metallic phase. The discussion in this chapter will deal specifically with the metallic component present on a nonreducible oxide, or perhaps carbon black, which is the support. The support is able to stabilize smaller metal particles than can be maintained by the pure metals. It may also contribute a catalytic function, such as acidity, or it may chemically modify the metals.

Characterization of the supported metal is essential to understanding its operation and for comparing performance in a meaningful way among catalysts. Catalyst purchasers specify values for known controllable physical parameters that affect the performance of the catalyst; they judge a manufacturer's catalyst against a list of specifications. Another major purpose for characterizing supported-metal catalysts is to establish a correlation between a structural feature and an aspect of catalyst performance. It might then be possible to optimize the catalyst by manipulating its structure. Of course, catalyst development over past decades has not had the benefit of any characterization except composition; however, the odds of achieving, through serendipity, the perfect tuning of a modern catalyst containing multiple metals, heteroatoms, and a structurally complex support are poor. This effort might be compared to attempting to tune a race car to win without knowing how it works.

Finally, metals characterization is a necessary preparation for diagnosing a catalyst failure. Valuable metals are included in catalyst formulations only as needed, without any "reserve" to sustain the catalyst if some of the metal is damaged—a catalyst vulnerability. The process restart might depend critically on an assessment of the present metal function compared to its state in a fresh, serviceable catalyst.

Characteristics of Supported Metals

Quantifiable characteristics of supported metals include particle size, surface area, dispersion, crystal structure, oxidation state, and purity as well as distributions of these values. Supported metals assume particle sizes and morphologies that are made possible only by the support. These are not trivial or simple concepts; studies have addressed the various definitions of particle size.[1] The dispersion is the fraction of total metal that is actually exposed at a surface. The support can stabilize oxidation states that otherwise would not persist under the reaction conditions. The extent of interaction of two or more metals in the same catalyst is extremely important, as the combined catalytic properties exceed what is possible with the separate unsupported metals.[2]

Conditions of Characterization of Supported Metals

A choice of conditions for characterization depends on the chemistry of the metal and the purpose of the study. If the chemistry or kinetics of the catalytic reaction is in question, then only the exact reaction conditions will do. The temperature, pressure, and composition of the feed and product streams are important. Multimetallic catalysts require special attention to conditions, since the surface composition of clusters containing more than one metal deviates from the bulk composition in a manner dependent on catalyst temperature and surrounding atmosphere.[3]

Most methods of characterization dictate the conditions of their application. Adsorption and chemisorption measurements, for example, are conducted in the range of pressures and at the temperature where the adsorption stoichiometry is established. Methods that rely on electron photoemission are usually conducted under high vacuum conditions to accommodate the radiation or to obtain useful electron paths. Also, electron microscopy is usually performed on samples in a vacuum. Commonly applied probes of metals in catalysts and the conditions of their use are listed in Table 2.1.

The suitability of characterization tools also depend on the objective of the characterization. Fingerprinting is the one of the least demanding purposes for catalyst characterization, since any method can be used in which the extrinsic variables that affect the result are recognized. Thus, even catalyst color might be a useful characteristic as long as it has no unrecognized dependence on humidity, for example, and as long as it actually correlates with other useful, but harder to measure, qualities.

2.2 Typical Approaches to Metals Characterization

Chemisorption

Chemisorption techniques are discussed at length in the lead volume in this series, *Encyclopedia of Materials Characterization*. Chemisorption is a simple and extremely informative, well-established method[5, 6] for metals characterization that involves measuring the quantity of gas required to form an adsorbed monolayer on a metal surface. The adsorbed gas is chemically bonded to only the metal's surface, which

Technique	Temperature	Gases	Pressure	Comment
Chemisorption	Usually 293 K	O_2, H_2, CO	Up to ≈ 30 kPa	
TPD	–	–	–	Parameters manipulated during measurement. Interpretation potentially complicated by changing conditions.
Probe reactions	–	–	–	Conditions defined by reaction and catalyst activity.
Electron microscopy	Variable	Vacuum	Vacuum	Environmental cells are under development. (See Ref. 4.)
X-ray diffraction	Variable	Variable	Variable	Suitable, but infrequently used for in situ studies.
X-ray absorption spectroscopy	Variable	Variable	Variable	Specifically suited for in situ studies.

Table 2.1 **Commonly applied probes of metals structure and catalyst environment conditions for their use. (The term "Variable" means the parameter is freely controllable and does not interact with the probe.)**

is different from the nonspecific Brunauer–Emmet–Teller (BET)[7] method for measuring surface areas, in which the gas is weakly and nonselectively adsorbed on all surfaces. Also, chemisorption may be used to measure dispersion, which is the ratio of surface metal atoms to total metal atoms in a sample. Surface atoms are assumed to be calculable from the number of adsorption sites, which can be quantified using chemisorption.

The success of the chemisorption method depends on knowledge of the conditions under which gases adsorb on metals with known stoichiometry. The stoichiometries of adsorption are determined most directly by measuring the adsorption on an unsupported (pure) metal with known surface area. The validity of the extrapolation to supported metals, which may consist of particles in the molecular size range, is tested by comparing adsorption uptake by the supported-metal catalyst to the metal surface area, which is estimated independently by X-ray diffraction, electron microscopy, or some other method. It is commonly observed that the actual adsorbate-to-metal ratio is close to unity for catalysts in which the average metal particle size is near 10 Å or smaller, in which case all the metal atoms are at the surface.

The most common gases used in chemisorption experiments are H_2, O_2, and CO. Hydrogen and oxygen dissociate on metals and are atomically adsorbed. Table 2.2 shows metals often used in supported-metal catalysts and gases that cannot be used for their characterization. Hydrogen is probably the preferred adsorbate in those cases in which it is adsorbed at all. Oxygen forms volatile oxides on Ru, Os, and Re, and it forms bulk oxides on several other metals. Carbon monoxide forms mobile carbonyl species on several metals and dissociates on a few. For some metals, the adsorption stoichiometry for hydrogen and CO exceeds unity for extremely small clusters because of multiple adsorption on the low-coordinate metal atoms that dominate at the surface. The accuracy of CO chemisorption measurements is the most often disputed.

When reviewing the numerous challenges in the literature to the suitability of chemisorption for metals characterization, one should keep in mind the broad base of data supporting the method. No other method offers the same simplicity, precision, and reproducibility. The value of a direct measurement of surface area cannot be overstated. The amount of gas adsorbed directly determines the number of surface atoms of the adsorber metal. It is often possible to select an adsorbent gas that is specific for only one of the metals in the catalyst.

Methods of Measuring Chemisorption

A chemisorption measurement requires only that the metal surfaces of a catalyst sample be free of contamination. This is accomplished by reducing the metal in H_2 and then evacuating the sample while it is still hot enough for the hydrogen to desorb rapidly from its metal surfaces. Finally, the catalyst sample is exposed to the chemisorbing gas under prescribed conditions, and the amount retained by the surface is measured.

Metal	**Adsorbate**	**Comments**
Fe	H_2, O_2, CO	
Co		
Ni	CO	The volatile nickel carbonyl may be formed.
Cu	H_2, O_2, CO	N_2O forms an oxygen monolayer at 90 °C and 200 torr, releasing one N_2 per oxygen atom.[8]
Ru	O_2	Ruthenium oxidizes to volatile RuO_4. CO may induce mobility and affect morphology.
Rh		CO adsorption stoichiometry approaches 2 as average particle size declines toward isolated Rh atoms.[9]
Pd		H_2 forms a hydride at room temperature and low pressure. Evacuation removes the absorbed hydrogen but leaves the surface (adsorbed) hydrogen.[10]
Ag	H_2, O_2, CO	H_2 and O_2 are not dissociated, and CO is not chemisorbed.
Re	H_2, O_2	H_2 uptakes are low for reasons that are not clear. Rhenium forms a volatile oxide.
Os	O_2	H_2 uptakes are low for reasons that are not clear. Osmium forms a volatile oxide.
Ir		Hydrogen and CO stoichiometries exceed unity on extremely small Ir clusters (< 10 Å).[11]
Pt		H_2, O_2, and CO chemisorptions for Pt characterization have broad acceptance.
Au	H_2, O_2, CO	H_2 and O_2 are not dissociated, and CO is not chemisorbed.

Table 2.2 Metals often used in catalysts, and gases that *cannot* be used for their characterization by chemisorption. The gases H_2, O_2, and CO are most commonly used.

Static volumetric Chemisorption   In this method, the pressure of a known volume of gas is measured; then the gas is expanded into an evacuated known volume that includes the catalyst, and the pressure is measured again. A static volumetric chemisorption unit is shown schematically in Figure 2.1. Several measurements are usually made with progressively higher pressures extending to about 30 kPa; the measured uptakes describe a curve called an isotherm. Some of the gas adsorbs weakly on the metals and support, so the catalyst is usually evacuated at the end of the first series of measurements, and the measurements are then repeated. The two isotherms that are generated with a 1% Pt/Al_2O_3 catalyst are shown in Figure 2.2. The upper trace is the total uptake of gas, which includes both strongly and reversibly absorbed hydrogen; the lower trace is the gas that was removed by evacuation in 15 min at room temperature and then replaced during the second

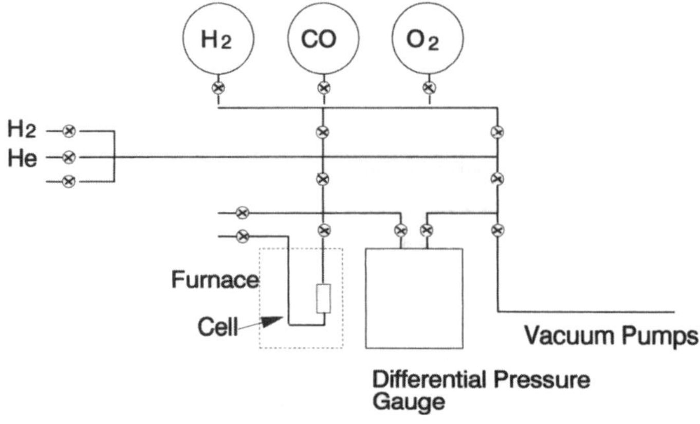

Figure 2.1   Static volumetric chemisorption apparatus.

isotherm. The actual dispersion of Pt in the catalyst is calculated by taking the difference between the values obtained by extrapolating the upper and lower curves back to zero pressure. This difference is the amount of gas that was strongly bound on the metal surface, on the basis of which the amount of surface metal was calculated. Unique values of adsorption uptakes are usually determined by extrapolating the isotherms back to zero pressure.

The static volumetric method has the advantage of high precision and high flexibility. In some cases, the amount of weakly adsorbed gas has been found also to correlate with some aspect of metal structure, for example, low-coordinate corner atoms that adsorb one atom strongly and a second one more weakly. The method's principal disadvantage is its slowness, since the catalyst and the gas phase must reach equilibrium at each measured point. Also, the apparatus and the measurements themselves are elaborate compared to the other common methods.

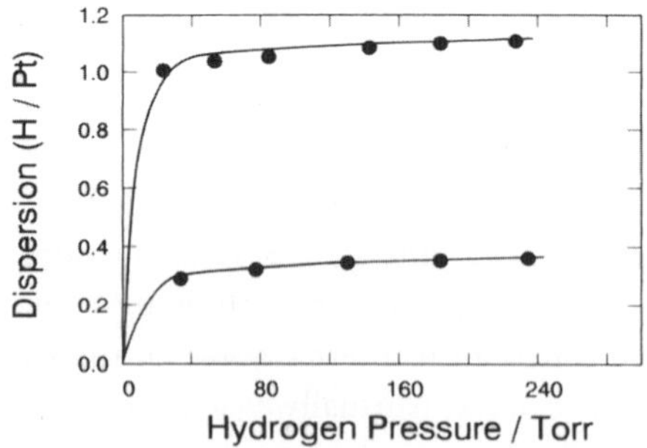

Figure 2.2   Adsorption isotherms for H₂ on 1% Pt/Al₂O₃. The upper trace was measured at room temperature after reduction and evacuation at 450 °C. The lower trace was measured after subsequent evacuation for 15 min at room temperature to about 1 × 10⁻⁵ torr. The vertical axis is the ratio of moles of atomic hydrogen to moles of platinum in the sample.

                                                    SUPPORTED METALS   Chapter 2

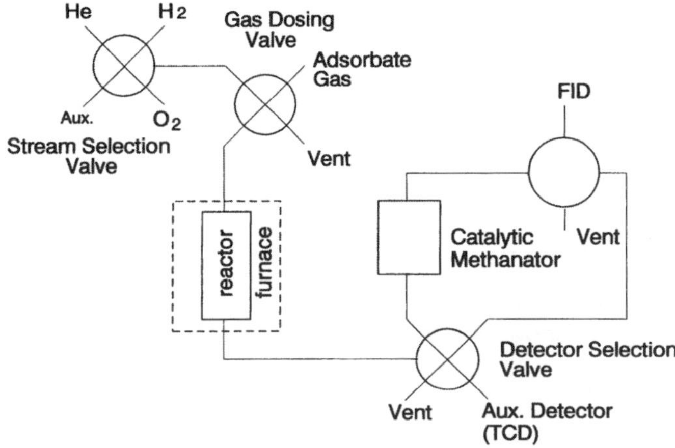

Figure 2.3 A versatile pulse chemisorption/temperature-programmed desorption apparatus. Pulses of adsorbate are injected into a carrier gas for pulse chemisorption. If CO is a desorbed molecule, it would be converted to methane and then analyzed by FID for about a 10^6 gain in sensitivity.

Pulse chemisorption   In the pulse method, a suitably reduced and clean catalyst sample is placed under a stream of inert gas, into which pulses of a known amount of adsorbing gas are injected. The amount of adsorbing gas that reaches the back of the reactor is monitored by a thermal conductivity detector or similar device. A pulse chemisorption apparatus is shown in Figure 2.3. (The same apparatus is suitable for temperature-programmed desorption and temperature-programmed reaction, which are described below.) The first few pulses may not emerge at all from the reactor; but when the catalyst is saturated, a pulse "breaks through," after which the amount of gas retained drops to a constant low fraction. The chemisorption uptake of the metal is obtained by integrating the amount of gas that was retained throughout the complete series of pulses. The amount of gas detected passing through the cell may never quite reach the amount delivered. This is because a small amount of gas is weakly adsorbed then released slowly between pulses, producing a uniform low background. Uptakes measured by pulse chemisorption correspond most closely to the strongly held fraction measured by the static volumetric method, though this depends on the system and conditions.

The pulse chemisorption method has the advantages of rapid turnaround and easy automation. Sensitive detectors have also made the method precise, though the retention and slow release of weakly adsorbed gas do present an ambiguity in the measured amount of adsorption.

Temperature-programmed desorption   The temperature-programmed desorption (TPD) and pulse chemisorption methods use the same apparatus; but the TPD method requires preparing a sample and deliberately exposing it to the adsorbate

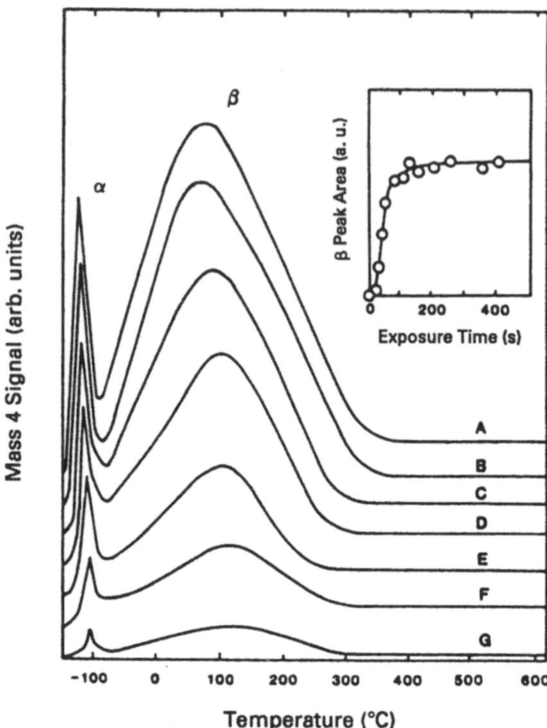

Figure 2.4 Temperature-programmed desorption (TPD) of D_2 from 1 wt % Pt/Al_2O_3. Adsorption pressure was 1×10^{-3} torr at −150 °C. Dose times were (A) 1000 s, (B) 260 s, (C) 80 s, (D) 50 s, (E) 40 s, (F) 20 s, and (G) 10 s. Inset shows the area of the β peak as a function of the dose time.[30]

gas to establish a saturated surface. The sample is then heated on a linear temperature program under a stream of inert gas. A thermal conductivity detector (TCD) or similar device monitors the amount of adsorbate leaving the reactor, which is equal to the saturation coverage of the sample. The temperature-programmed desorption spectrum for the desorption of deuterium from 1 wt % Pt/Al_2O_3 is shown in Figure 2.4. In general, the temperature of the desorption reflects the strength of the adsorption; the weakly bound fraction of adsorbate is separated in temperature from the strongly bound fraction. Thus, the TPD method combines the strengths of the static volumetric and pulse methods. The desorption peak shapes also contain kinetic information on the desorption process. This information may be difficult to interpret, since it is collected under conditions of changing temperature, coverage, and partial pressure of the desorbing species. In particular, readsorption of the desorbate complicates peak shapes.

Temperature-programmed reaction (TPR) may be regarded as a variant of TPD in metals characterization. For example, methanation of CO by some metals strengthens the TPD experiment. In this case, the catalyst, with its surface saturated

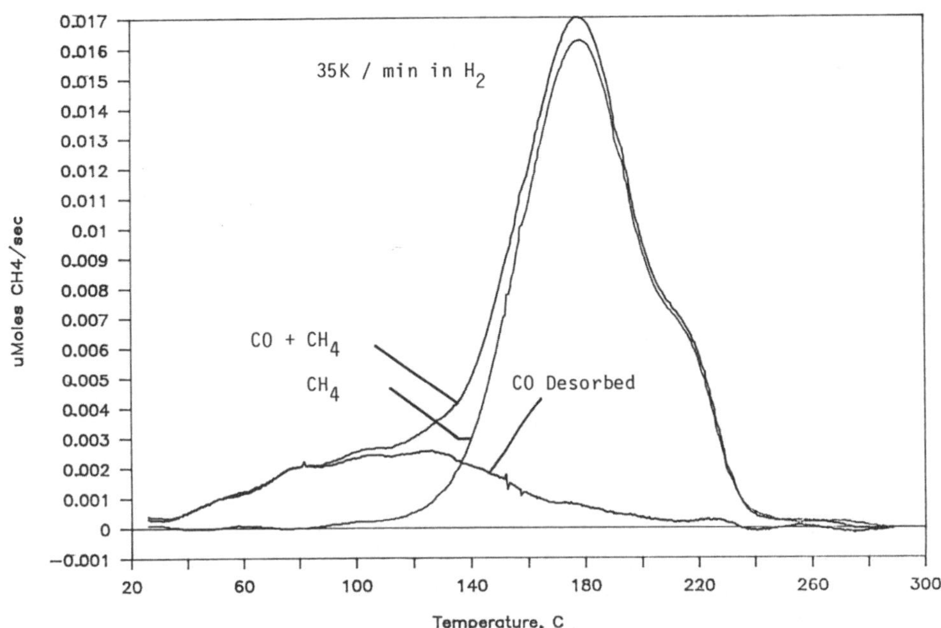

Figure 2.5   **Temperature-programmed desorption/reaction (TPD/TPR) of CO from a supported group VIII metal. The desorption was in H_2 with a temperature ramp rate of 35 K/min. The CO + CH_4 curve is the FID signal reflecting all desorbed products after treatment in methanator.[31]**

by CO, is heated under a stream that contains H_2. Then, CH_4 released by the surface can be monitored with a flame ionization detector (FID) for an improvement in sensitivity over the TCD. Figure 2.5 shows traces representing reaction and desorption of CO from a supported group VIII metal catalyst. Note that some CO desorbed below the methanation temperature. All the CO was measured, whether methanated by the cobalt or not, by passage with H_2 through a special methanation reactor. An advantage of the TPR method is that the CH_4 is inert compared to CO; it does not readsorb during the experiment, thus preventing the distortion of desorption traces. In any case, it is usually advantageous when measuring CO TPD to pass the desorbed CO, with H_2, through a methanation catalyst so the product can be measured by an FID.

2.3   Reaction Studies of Supported Metals

Studies of catalysts based on measuring well-understood probe reactions complement chemisorption studies in that both methods measure surface sites. Reaction studies of catalyst structures have the conspicuous advantage of providing information directly on active catalytic sites; these could be the only studies that matter in the event that active sites comprise a tiny fraction of the catalyst surface. Probe reactions can be categorized as structure insensitive or structure sensitive.[12]

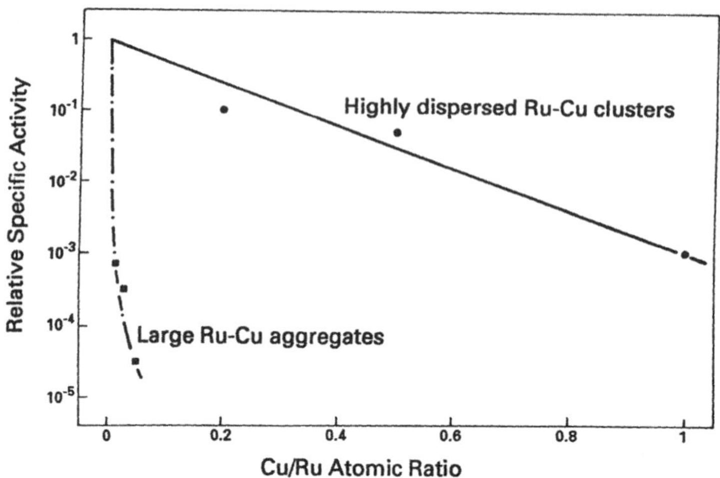

Figure 2.6   Relative rates of ethane hydrogenolysis catalyzed by Ru–Cu bimetallic catalysts.[32] The specific rates are based on sites measured by hydrogen chemisorption. The large Ru–Cu aggregates and highly dispersed clusters had average diameters of about 100 Å and 10 Å, respectively.

Hydrogenation of ethylene is structure insensitive—at least it is on catalysts having either metal clusters as small as several atoms or an active metal intermixed with an inactive metal.

Structure-sensitive reactions, in contrast, require a complex catalytic site, which is easily disrupted by surface contamination, or surface steps and kinks that comprise the surfaces of very small clusters. For example, ethane hydrogenolysis is structure sensitive.[13] A mixture of copper, which is inactive for hydrogenolysis, with ruthenium, which is active, provides a clear example of structure sensitivity. A small atomic fraction of copper added to ruthenium has been observed to decrease the activity of poorly dispersed ruthenium for ethane hydrogenolysis by five orders of magnitude, but the same amount of copper had much less effect on highly dispersed ruthenium (see Figure 2.6). This effect was not the result of a loss of surface area—the metal's surface area actually increased as the catalytic activity declined. The copper, which is immiscible in ruthenium, had spread onto the ruthenium and partially coated the ruthenium surface.[14] The magnitudes of difference in the effect of copper on highly dispersed and poorly dispersed ruthenium indicated that the copper was segregated at the ruthenium surface. The highly dispersed ruthenium presented a much greater surface area and required more copper than the poorly dispersed ruthenium to achieve a given level of contamination. The extreme sensitivity of the hydrogenolysis reaction to copper content has been interpreted as evidence that the reaction requires as many as eight contiguous ruthenium atoms. The concept of the ethane "landing site" on metal surfaces has survived, and ethane continues to be used to inspect the composition of mixed metal surfaces.

                                                SUPPORTED METALS   Chapter 2

2.4 X-ray Diffraction and Scattering Methods

These techniques are discussed at length in the *Encyclopedia of Materials Characterization*. X-ray diffraction methods of catalyst characterization are general in nature and very powerful, and, with chemisorption, they are probably most widely applied to metal phases in supported-metal catalysts. Diffraction is a property of crystals that are physically large compared to the wavelength of X-rays. A metal's diffraction pattern identifies the packing structure of its metal phase, provided the metal is crystalline, is in sufficiently large particles, and has a diffraction pattern that cannot be obscured by the support. Supported single metals adopt their native packing when they form crystallites sufficiently large to define a structure; the situation is more complex when there are two metals. In those cases, one or more mixed phases may form, or one metal may be crystalline and the other not, or one or both metals may adopt non-characteristic packing.

Particle Sizes from Line-Broadening

Metal clusters of the approximate size of the wavelength of X-rays (e.g., 1.54 Å from a laboratory copper anode source) are the most interesting to catalysis. As the particles get small compared to the X-ray wavelength, their diffraction gets increasingly diffuse. This line-broadening is exploited for determination of metal particle size.[15] The relationship of crystallite size to X-ray line-broadening is given by the Scherrer equation:

$$D = \frac{K\lambda}{\beta \cos\theta} \tag{2.1}$$

where D is a particle diameter, λ is the X-ray wavelength, and θ is the Bragg angle. The variable β is the pure diffraction line width, which is corrected for K_α X-ray doublet separation and instrumental broadening, and is expressed in radians in terms of 2θ. The constant K depends on the particle shape and the line indices. Values given to K range from about 0.7 to 1.7, but it seems a reasonable practice to let it equal 1.0 so that adjustments are made from a consistent starting point.

In its simplest application, β is the full line-width at half the maximum intensity. The instrumental broadening is determined by measuring a line-width from a massive annealed sample of the metal, which gives the line-broadening. Determination of instrumental broadening is not simple; disorder is surprisingly high in unannealed metal foils and powders. The broadened line is corrected for instrumental broadening by either Equation 2.2 or 2.3.

$$B = \beta + b \tag{2.2}$$

$$B^2 = \beta^2 + b^2 \tag{2.3}$$

In these equations, the variable B is the measured line-width and b is the instrumental broadening.

Widths of X-ray diffraction lines are affected by internal disorder, compositional inhomogeneity, and strain, as well as particle size. These effects can be separated, and the apparent particle size distribution can be derived by the Fourier analysis of diffraction line shapes.[12] The Fourier analysis is considerably more involved, requiring cleaner data than the simple line-width calculation. Fourier analysis would seem hardly worthwhile if the sample were not reduced in situ to reverse strain and corrosion induced by oxidation.

Particle size estimates from line-broadening have a high degree of uncertainty, given that line-width is not a very precise quantity and that particle shape and disorder are usually ignored. The transitional aluminum oxides have diffraction lines that interfere with measurements of line-broadening from $4d$ and $5d$ metals. However, the line-broadening technique is fast and easy, and its results are not appreciably influenced by chemical effects that could interfere with chemisorption measurements. There exists a large data base for line-width comparison to determine particle size.

Small-Angle X-ray Scattering

Small angle X-ray scattering (SAXS) (see Brumberger[16]) is currently a seldomused technique. Like the other X-ray techniques, SAXS is compatible with in situ studies. It characterizes all the metal in the sample. Particle size information from SAXS depends only on the average physical dimension of regions of uniform electron density, not on the crystal packing or presence of strains in these regions.

As it is most commonly practiced, SAXS interpretation is based on the Law of Guinier:

$$I(s) = (\Delta\rho)^2 \exp(-4\pi s^2 R_g^2) \tag{2.4}$$

where s is the wave vector $s = [4\pi\,(\sin\,\theta)/\lambda]$, R_g is the radius of gyration, and $\Delta\rho$ is the difference in electron density across an interface. Equation 2.4 predicts that a plot of the logarithm of the small-angle intensity versus the square of the scattering angle should yield a straight line with a slope proportional to R_g^2. It is important to recognize that this equation is based on assumptions that make it valid only at very small angles. The radius of gyration is the radius of a sphere with a mass equivalent to the scattering particle and with equal scattering properties.

As stated, SAXS provides values for the mean dimensions of regions with internally-uniform electron densities. Its usefulness in the study of supported metals is complicated by the catalysts actually having three phases—the support, the metal, and the voids and porosity of the support. Voids and porosity can be eliminated as a source of scattering by filling them with a liquid, often $C_2H_4Br_2$ or CH_2I_2, that has an electron density that matches the support. An alternative analytical method for deconvoluting the porosity contribution to the small angle scattering is described by Boudart and Djega-Mariadassou.[12]

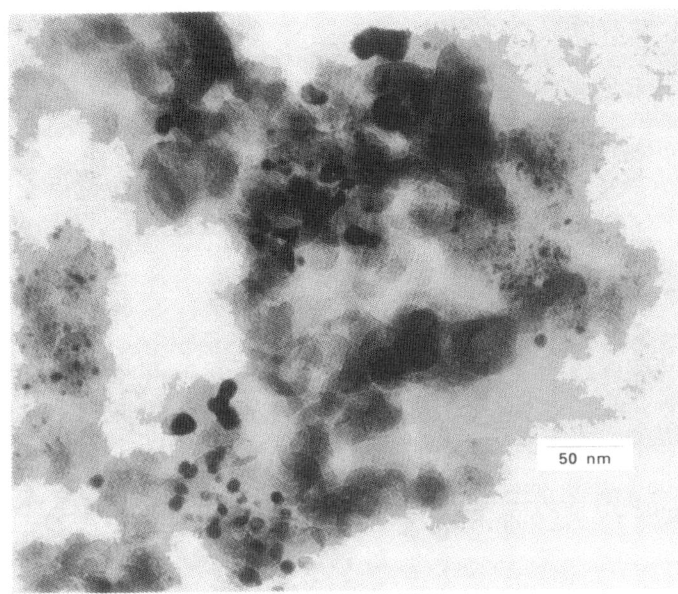

Figure 2.7 Bright field TEM image of 5 wt % Pt–3 wt % Sn/SiO$_2$ dried in air at 380 K and then reduced under hydrogen at 775 K.[33]

2.5 Electron Microscopy

Transmission electron microscopy and scanning transmission electron microscopy are discussed at length in the *Encyclopedia of Materials Characterization*. Metals characterization with electron microscopy[17] is uniquely satisfying in that one can actually see the metal particles. The appearance of these particles is, however, somewhat illusory, since contrast is generated by different mechanisms in electron micrographs than in optical images. As one colleague put it, "Not everything you see in an electron micrograph is an artifact." An experienced microscopist is invaluable for interpreting electron micrographs.

Scanning transmission electron microscopy (STEM) is the best technique for directly determining a wide range of metal particle sizes. It is routinely used to image metal clusters with diameters below 1 nm. Conventional transmission electron microscopy (TEM) can image particles larger than 1 nm and identify the composition and structure (single crystal or polycrystalline) for larger clusters. Figure 2.7 is a TEM image of 5 wt % Pt–3 wt % Sn/SiO$_2$ reduced in H$_2$ at 500 °C. Figure 2.8 is a microdiffraction pattern of a single 20 nm particle in the same sample, as measured in the TEM. The pattern indicates that the particle had a twinned, ordered Pt$_3$Sn structure.

Metal cluster size information from electron microscopy is usually presented in particle size histograms, which plot actual numbers of particles in specific ranges of

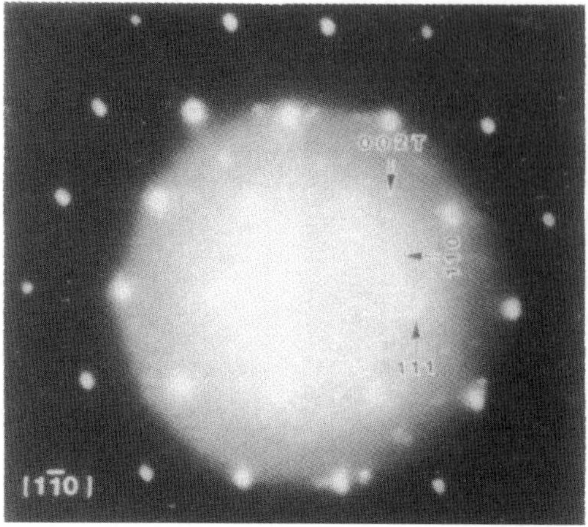

Figure 2.8　Microdiffraction pattern obtained from a single ~20 nm PtSn particle in Figure 2.7 showing twinning and an ordered Pt_3Sn structure.[33]

particle sizes. Computerized image processing eliminates the tedium of manually counting and measuring hundreds of tiny, indistinct spots on the image. Figure 2.9 is a set of histograms showing particle size distributions for a Pt/Al_2O_3 catalyst treated at successively higher calcination temperatures.

Another operating mode of electron microscopy is called annular dark-field imaging (sometimes simply called dark-field), in which images are generated from particle diffraction. The imaging device is displaced by an angle to measure the Bragg-diffracted beams from individual crystalline clusters. The detector can be placed where only the crystallites of a selected structure contribute to the image.

The necessity that a sample be maintained under vacuum for electron microscopic imaging imposes an important limitation on the ability of this technique to characterize catalysts in their active state; however, progress is being made in developing sample chambers that maintain a useful gas pressure over the sample during imaging. The biggest disadvantage of electron microscopy may be the very small sampling region of the catalyst that is actually studied. The utility of the technique depends on the skill of the microscopist for imaging work that provides a true sampling of the catalyst and not just images of the most photogenic regions.

2.6　X-ray Absorption Spectroscopy

This method is discussed at length in the *Encyclopedia of Materials Characterization;* also, literature is available that covers all aspects of X-ray absorption spectroscopy.[18, 19] X-ray absorption spectroscopy (XAS) is a general term that includes

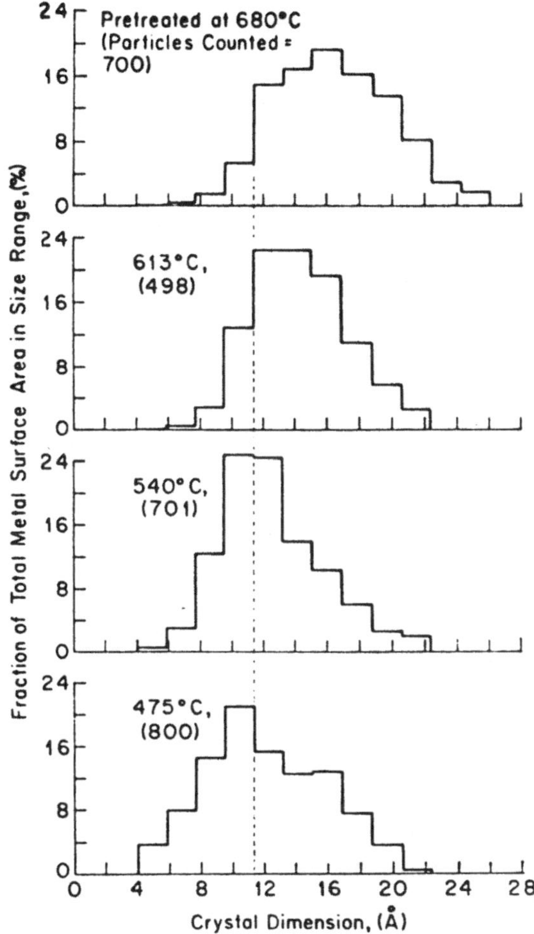

Figure 2.9 Platinum particle size distributions from TEM images of 1 wt % Pt/Al$_2$O$_3$ reduced in H$_2$ at 500 °C and then calcined at the temperatures shown in the figure.[34]

X-ray absorption near-edge structure spectroscopy (XANES) and extended X-ray absorption fine-structure spectroscopy (EXAFS). The XAS methods are almost unique in that they can provide electronic and physical structural information on a catalyst while it is catalyzing a reaction—in principle, under any conditions. This is because the spectra are measured with energetic "hard" X-rays with high penetrating ability. Thick (strong) windows, gases at high pressure, and liquid feeds will not usually interfere with the measurement of spectra. Moreover, XAS is element specific, which helps in the study of metals that comprise much less than 1 wt % of the catalyst. The information contained in X-ray absorption spectra can include chemical species and the oxidation state of the absorber, average numbers and types

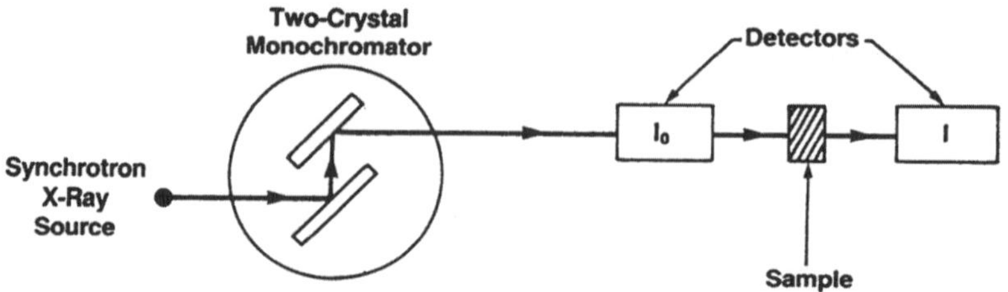

Figure 2.10 Setup for transmission X-ray absorption spectroscopy. Synchrotron X-ray sources provide broad continuous spectra with high intensity in the energy range from below 100 eV to above 30 keV.

of atoms within about 5 Å of the absorbing element atoms, average distances to these atoms, and the distributions within these distances.

Preparation for Measurements

X-ray source and spectrometer   The measurement of X-ray absorption spectra requires a high intensity X-ray source that provides radiation over a wide energy range. It was shown in the 1970s that synchrotron radiation made routine XAS feasible; in fact, most measurements of X-ray absorption spectra are made at synchrotron light sources. Many sources now exist that are dedicated to the production of X-rays for experiments such as XAS and diffraction.

The X-ray source apparatus for XAS is a beamline—a pipe leading tangentially off a curved section of a synchrotron so that the synchrotron radiation passes down it. An experimental setup for XAS is diagrammed in Figure 2.10. Most monochromators are two-crystal or channel-cut crystal devices. Many detectors are available, including ion chambers, proportional counters, scintillation counters, and solid state detectors. Fluorescence detection is preferred for extremely dilute samples, and electron yield or reflection geometry detection schemes provide surface sensitivity.

Choice of edges   Hard X-ray spectra result from the excitation of core electrons. The K and L_{III} edges are strong enough, corresponding to $1s$ and $2p$ electron excitations, respectively. The choice between K and L edges depends mainly on the overlap between the accessible energy range of the beamline and the binding energies of the $1s$ or $2p$ electrons. Figure 2.11 illustrates this point. In general, the K edge is used to measure the spectra of $3d$ and $4d$ metals, and the L_{III} edge is used to measure $5d$ metal spectra. Either type of spectrum provides EXAFS that yields the same physical structural information, but the information obtained from L_{III} near-edge spectra is slightly different than that from K near-edge spectra.

Interpretation of Information

LIII near-edge structure   The L_{III} X-ray absorption spectrum from platinum in the form of a 7.5 μm foil is shown in field A of Figure 2.12. The edge is the sharp

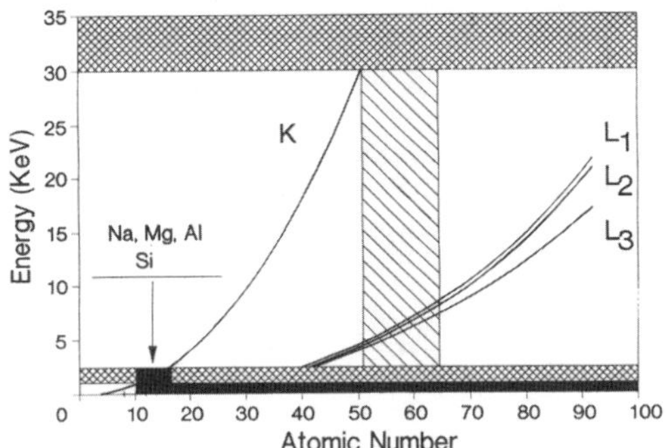

Figure 2.11   **Energies of K and L absorption edges. The ranges from about 1 keV to 2 keV and above 30 KeV are problematic due to source and optics limitations. At lower energies the small separation of the L edges interferes with interpretation of near-edge structure or EXAFS. Thus, there exist ranges of elements—sodium through silicon and from around tin through the lanthanides—that are difficult to study using XAS.**

increase in absorbance that results from the onset of excitation of Pt $2p$ electrons with binding energy of 11.6 keV. The near-edge spectrum refers to a region about 200 eV wide centered about the edge itself. The important information is the intensity or area of the edge resonance (or white line) just above the edge. The edge results from the excitation of the $2p$ electrons into the d band; it varies in intensity with the availability of vacancies in the d band. Accordingly, the edge resonance is sensitive to the oxidation state of the absorbing metal. Figure 2.13 shows how the L_{III} near-edge spectra of iridium in 1% Ir/Al_2O_3 were used to monitor the reduction of the iridium as it was heated in H_2. The average oxidation state of the iridium has an almost linear dependence on the intensity of the edge resonance.

K near-edge structure   Several complications can interfere with the measurement or interpretation of L_{III} edges of the $3d$ and $4d$ metals. Measurements of K-edge spectra should then be taken instead. The dipole allowed-transition takes $1s$ electrons to p-type orbitals, which are filled in the transition metals. Therefore, the white line is absent from K-edge spectra. However, distortions of orbitals or hybridization that mixes s, p, and d orbitals make K-edge spectra very sensitive to the chemical bonding and bonding geometry of the absorber; thus, the spectra are useful as chemical fingerprints. These edges still contain direct information on the oxidation state in the form of binding energy shifts that correlate with oxidation states. Figure 2.14 shows that the molybdenum K edge, at 19 999 eV for the metal, was shifted for Mo in different compounds.[20] The energy shifts in Figure 2.14 are plotted against "coordination charge" on the molybdenum, which is calculated from formal oxidation states, the number of ligand bonds, and ligand bond covalencies. The correlation of edge shifts would obviously not be so strong with formal oxidation states, which are idealizations.

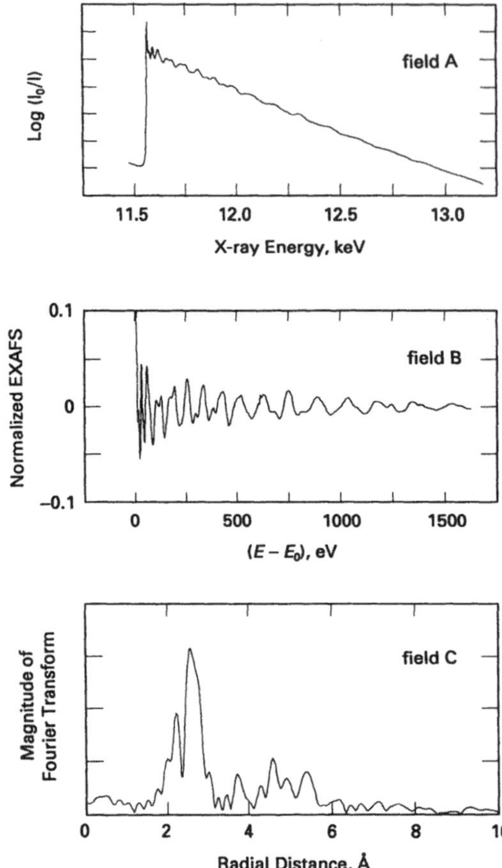

Figure 2.12   Field A: X-ray absorption spectrum of platinum in a 7.5 µm thick Pt foil; field B: platinum EXAFS is shown after background subtraction; field C: magnitude of the Fourier transform of the EXAFS shown in field B.

Extended X-ray absorption fine-structure   The EXAFS region of the X-ray absorption spectrum provides a description of the average physical environment of absorbing element atoms. The EXAFS is the oscillation in absorption that may extend more than 1000 eV beyond an absorption edge. The EXAFS of the platinum in the foil is visible in the spectrum in field A of Figure 2.12 and is shown after subtraction of the background in field B. The EXAFS is essentially a scattering pattern produced in the sample by the photoejected electron as its kinetic energy increases. The EXAFS is described by Equation 2.5.

$$X(K) = \sum_{j} \sum_{i} \frac{N_{ij}}{KR_{ij}^2} F_i(K) \exp(-2K^2\sigma_{ij}^2) \sin\left[2KR_{ij} + \phi(K)\right] \tag{2.5}$$

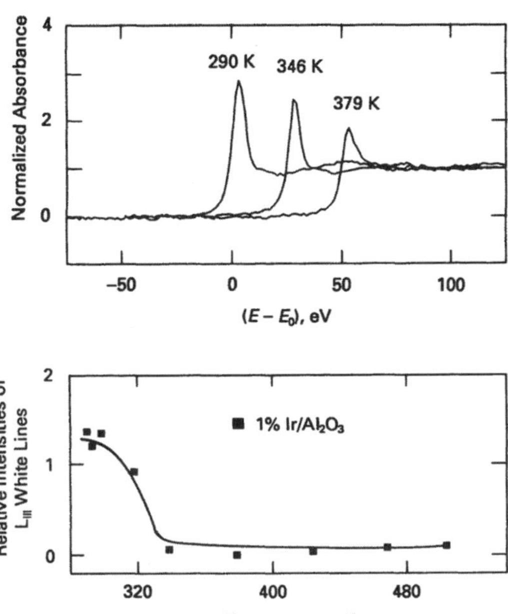

Figure 2.13 The reduction in hydrogen of 1%Ir/Al$_2$O$_3$ as monitored by measuring the iridium white-line intensity during heating. The upper field shows changes in the white-line during reduction. The lower field shows the intensity changing as a function of temperature.[35]

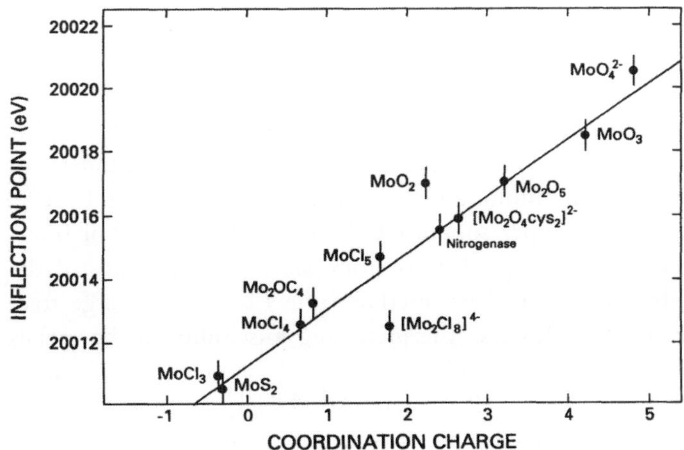

Figure 2.14 Shifts relative to the pure metal in Mo K edge positions, and in corresponding binding energies of 1s electrons, are plotted for a number of Mo compounds.[20]

where N is the number of atoms at distance R_{ij} of type i, and K is the electron wave vector defined by

$$K = \frac{2\left[4\pi^2 m^2 (E - E_0)\right]^{1/2}}{h} \tag{2.6}$$

where E and E_0 are the photon energy and the electron binding energy, respectively, m is the electron mass, and h is Planck's constant. The functions $F_i(K)$ and $\phi_i(K)$ in Equation 2.5 are the electron scattering factor and phase shift, respectively. The electron scattering factor is a characteristic of the scattering element, and the phase shift is a characteristic of the absorber–scatterer pair. The σ^2 is the mean-square relative displacement within the interatomic distance R. One sum is taken for all types of atoms within the coordination sphere of the average absorbing atom, and the other sum is taken for all coordination spheres.

The Fourier transformation of the EXAFS, shown in field C of Figure 2.12, produces a distorted radial distribution function called a radial structure function (RSF) to distinguish it from the real radial distribution function. The distortion from $F(K)$ and $\phi(K)$ identifies the scattering element. In Figure 2.12 field C, peaks indicating the radial positions of coordination spheres of platinum atoms appear at about 2.5 Å and in the range of 3.5–5 Å. The splitting evident in the 2.5 Å peak is due to the functional forms of $F(K)$ and $\phi(K)$ and is characteristic of platinum. All the peaks are shifted to positions short of the true radii by the phase function $\phi(K)$. The magnitudes of the peaks depend mainly on the number of atoms in the coordination spheres, the radii of the spheres, and the mean-squared relative displacements about the radii. Actual EXAFS analysis usually involves fitting a function, calculated from Equation 2.5, to the Fourier-filtered EXAFS from the first coordination shell.

Strengths and Weaknesses of XAS

X-ray absorption spectra provide information that does not depend on empirical analysis. Figure 2.15 shows the compositions of the first coordination shells of rhenium, osmium, iridium, platinum and gold in samples containing 1 wt % of the $5d$ metal and 0.3 wt % copper supported on Al_2O_3. The extent of interaction between the metals in each pair departs from the bulk miscibility of the metals because the number of "surface" or interface atoms becomes a large fraction of all atoms in the smallest clusters. The increasing miscibility of the metals from left to right in the figure is indicated by increasing interaction. Rhenium and copper are immiscible, but the small clusters placed every atom at an interface and thereby enabled a high degree of interaction. Interferences are rare in XAS because the X-rays are very penetrating, and only one element usually absorbs in the region covered by a given spectrum, unless the sample contains two elements that are adjacent in the periodic chart. Also, XAS is almost insensitive to sample

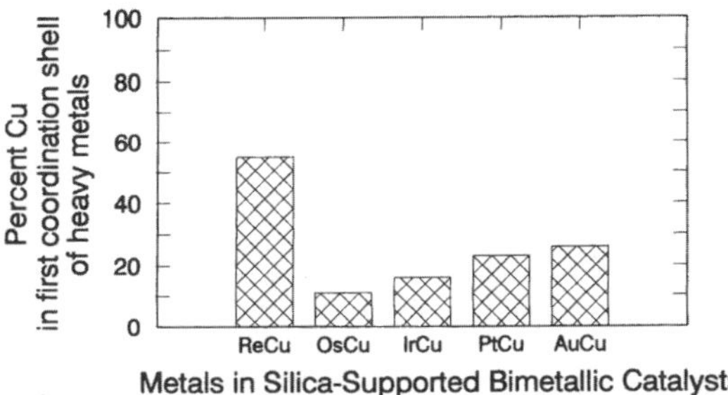

Figure 2.15   Compositions of the first coordination shells of the 5*d* metals in silica-supported samples each containing 1 wt % 5*d* metal and about 0.3 wt % copper.[36]

temperature, any gases, reaction feeds, supports, or other materials present with the metals in the cell.

The greatest weakness of XAS is probably the requirement of a synchrotron radiation source in order to measure spectra. Access to these is somewhat limited, although new synchrotron sources are being built. It is difficult to spend enough time at these sources to first gain proficiency and then measure the needed spectra.

2.7   Mössbauer Spectroscopy

The theory and technology of Mössbauer spectroscopy are described in Herber[21] and Greenwood and Gibb.[22] Specific applications of Mössbauer spectroscopy to heterogeneous catalysts have also been reviewed.[23] The Mössbauer effect is the resonance fluorescence of nuclear gamma radiation; it is observed in solids during the recoilless emission and absorption of the radiation. It is used in a spectroscopic technique, in which the source is an excited nucleus that relaxes with the emission of a gamma ray.[17] The difference in nuclear environments between the source and the absorber is measured as a doppler shift, typically within the range -10 mm·s^{-1} to 10 mm·s^{-1}, that is imparted on the gamma photon with respect to the sample nucleus.

The Mössbauer spectrum of atoms in a single phase is sensitive to the structure of the phase. Hence, these spectra provide specific fingerprints, linear combinations of which are used to determine the distribution of an element among different phases. If more than one valence state of an element is present, a semiquantitative determination of each state is possible. Figure 2.16 shows Mössbauer spectra for iron-exchanged Linde-type Y-zeolites versus a ^{57}Co–Cr source.[24] Both traces were interpreted as being due to Fe^{2+} with a small amount of Fe^{3+} impurity. The upper trace is from a sample saturated with water at room temperature; the lower trace is from the same sample after evacuation at room temperature for 24 h. Note that the recoil-free fraction gave a greater intensity after the evacuation. This means that the Fe^{2+} was essentially in solution before evacuation, but was localized on specific binding sites afterward.

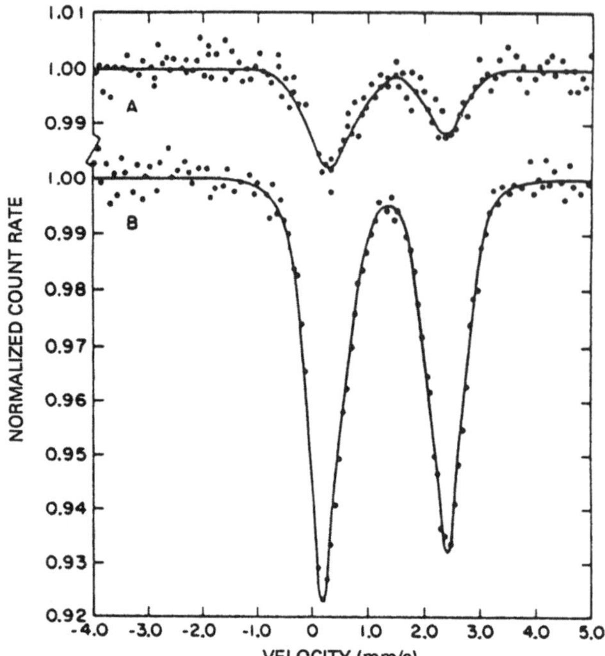

Figure 2.16　Representative Mössbauer spectra for iron-exchanged Linde-type Y-zeolite versus ^{57}Co–Cr source: (A) absorber saturated with water vapor at room temperature; (B) absorber evacuated at room temperature for 24 hours to a final pressure of 10^{-5} torr.[24]

Only elements for which there exists a suitable Mössbauer nuclide may be used for Mössbauer spectroscopy. The nuclides are listed in Table 2.3. Research is currently under way to make a synchrotron (tunable) gamma source, which will extend Mössbauer to all the elements. The technique can be used for many types of in situ studies; however, the recoil-free fraction of emission or absorption events declines as the source and sample are heated. For this reason, it is often necessary to cool them both, in some cases to a liquid helium temperature. Still, the gamma rays are penetrating and are insensitive to the composition and pressure of the atmosphere over the sample and the to the presence of other elements within the sample.

2.8　Photoelectron/Photoemission Spectroscopy

These techniques are discussed at length in the *Encyclopedia of Materials Characterization*. Photoelectron spectroscopy is used to measure core level (X-ray photoelectron) or valence level (ultraviolet photoelectron) binding energies. These two types of photoelectron spectroscopy differ both in the energy of the radiation and in the binding energies of the correspondingly excited electrons. They can provide

　　　SUPPORTED METALS　Chapter 2

Nuclide	E_γ, keV	$t_{1/2}$, s $\times 10^{-8}$	Nuclide	E_γ, keV	$t_{1/2}$, s $\times 10^{-8}$
^{40}K	29.4	0.39	^{160}Dy	86.8	0.205
^{57}Fe	14.38	9.8	^{161}Dy	25.6	2.8
^{61}Ni	67.4	0.53		74.5	0.30
^{67}Zn	93	940	^{166}Er	80.6	0.183
^{83}Kr	9.3	14.7	^{169}Tm	8.41	0.39
^{99}Ru	90	2.0	^{170}Yb	84.2	0.161
^{107}Ag	93	4.4×10^9	^{171}Yb	66.7	[0.05]
^{119}Sn	23.8	1.85	^{177}Hf	113.0	0.052
^{121}Sb	37.2	0.35	^{181}Ta	6.25	680
^{125}Te	35.6	0.14	^{182}W	100.1	0.14
^{127}I	59	0.18	^{183}W	46.5	0.015
^{129}I	26.8	1.63		99.1	0.052
^{129}Xe	40	0.096	^{186}Os	137.2	0.084
^{131}Xe	80.2	0.050	^{187}Re	134.2	0.0010
^{133}Cs	81	0.623	^{188}Os	155.0	0.72
^{141}Pr	145	0.2	^{191}Ir	129.4	0.0131
^{149}Sm	22	0.76	^{193}Ir	73	0.60
^{151}Eu	21.6	0.88	^{195}Pt	98.9	0.017
^{153}Eu	97.5	0.014		129.7	0.055
	103.2	0.38	^{197}Au	77.3	0.18
^{155}Gd	86.5	0.586	^{237}Np	59.6	6.3
^{159}Tb	58.0	0.013			

Table 2.3 Mössbauer nuclides (where the Mössbauer effect has actually been seen).[21] The E_γ and the $t_{1/2}$ are the energy of the Mössbauer gamma radiation and the half-life of the isotope, respectively.

a chemical analysis of the surface from assignment of the measured binding energies and also a determination of atomic charges (oxidation states) from the "chemical shifts" of the binding energies. The experiment is simple: electrons with binding energies E_b are photoejected with kinetic energy E_{kin} by interaction with photons of frequency v. The binding energies are then determined as follows:

$$E_b = h v - E_{kin}$$

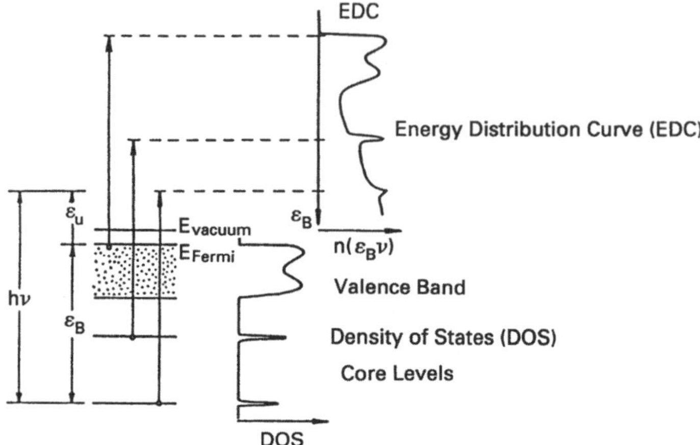

Figure 2.17   Relationship between the true density of states (DOS) of a material and the energy distribution curve (EDC) measured by photoemission spectroscopy.[25]

Vacuum ultraviolet is by definition radiation that interacts with valence electrons, whereas X-rays interact with core electrons. Figure 2.17 shows the relationship of the actual density of states versus energy on an atom and the energy distribution curve that might be measured by X-ray and UV photoemission.[25] The sampling depth depends on E_{kin} in a way described by mean free-path curves. In general, ultraviolet photoemission spectroscopy samples to a greater depth than X-ray photoemission, since the UV photoelectrons have lower kinetic energies.

Figure 2.18 is a diagram of an experimental setup, which is basically the same for UV and X-ray photoemission. The UV source is usually a glow discharge; the X-ray source is usually an aluminum- or magnesium-anode X-ray tube. Synchrotron radiation offers advantages for both these spectroscopies. First, synchrotron sources provide both high intensity and high resolution by virtue of their brightness. Also, the tunability of the source provides these advantages:

- Variable source energy helps to separate photoemission lines (varying in energy with the energy of the exciting source) from Auger lines (constant energy).

- The energy of the exciting source can be tuned to increase the absorption cross-section of the transition of interest.

- The energy of the source can be tuned in order to vary the E_{kin}, thereby varying the depth sensitivity of the measurements.

In one set of experiments, Pt clusters of uniform size, with one to six atoms per cluster, were mass-selected before their deposition onto a silica substrate. Photoemission studies were made of the Pt $4f_{7/2}$ and Pt valence band ($5d$) binding energies of the mass-selected clusters.[26] Figure 2.19 shows plots of Pt $4f$ and valence

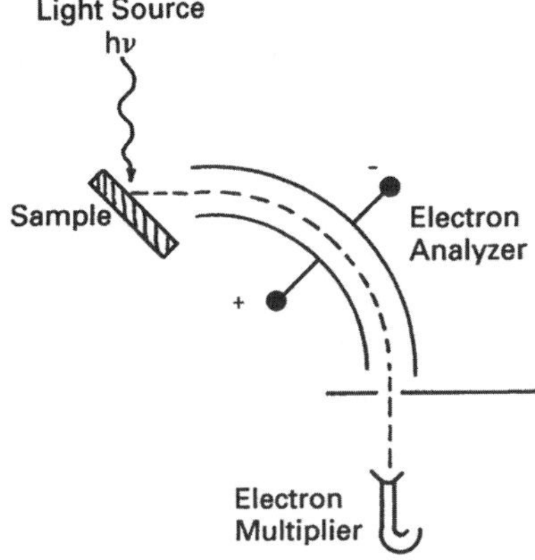

Figure 2.18   A photoemission spectrometer, which is entirely enclosed in a UHV chamber. The setup is essentially the same for UPS and for XPS, except for the radiation source.

band binding energies as functions of Pt atoms per cluster. The photoemission study indicated that these small Pt clusters exhibited a band gap and were therefore nonmetallic; it also indicated that the electronic structures differed for each size of cluster.

Photoemission techniques are most often useful for determining the oxidation states of metals in catalysts. Changes in oxidation states are realized as shifts in core or valence electron binding energies. Because of this, the applicability of photoemission to supported metals is limited, since the metals are almost always mounted on an insulating refractory oxide. The charging of the metal clusters that develops

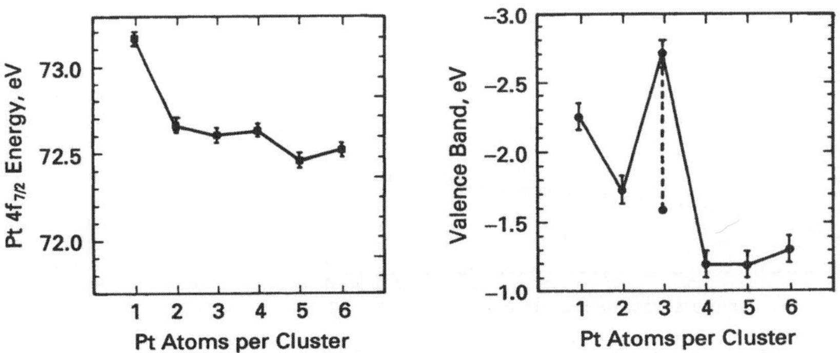

Figure 2.19   Binding energies of Pt $4f_{7/2}$ electrons and of Pt valence band electrons, measured by XPS (left plot) and UPS (right plot). The valence band binding energy for three atom clusters had two values. This was believed to indicate the coexistence of linear and ring type clusters.[26]

when photoemission occurs broadens and shifts the apparent binding energies and interferes with the determination of oxidation states.[27] Fortunately, strategies do exist to deal with the charging problem.

2.9 Magnetic Methods

Applications of magnetic methods to catalytic materials, which are limited to ferromagnetic metals in catalysts, are reviewed by Selwood.[28] Measurements of magnetization have been used to study particle size—by direct means and by measuring

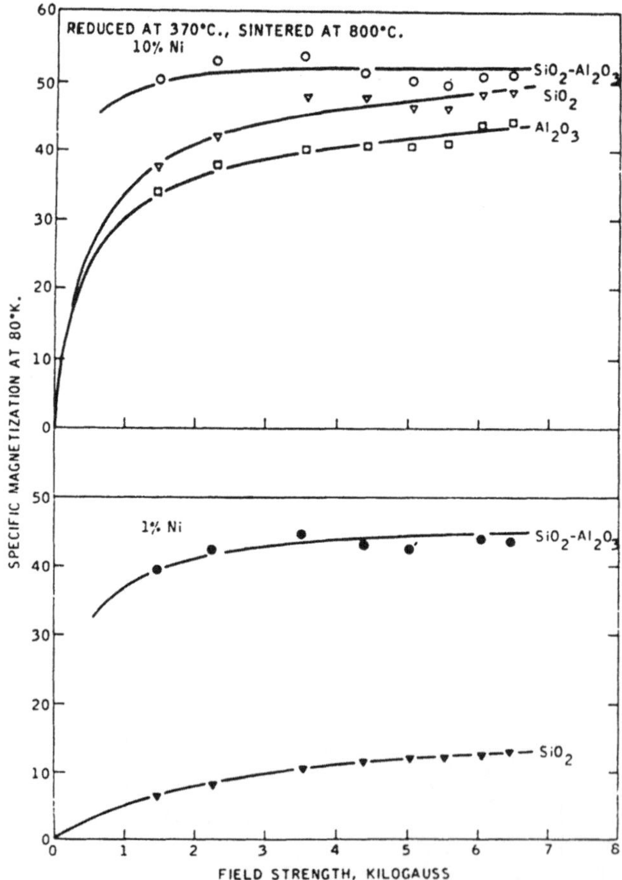

Figure 2.20 Specific magnetization curves for 1 wt % Ni supported on SiO$_2$, Al$_2$O$_3$, and SiO$_2$–Al$_2$O$_3$.[28] The low specific magnetization of nickel supported on SiO$_2$ was a result of its high dispersion, and not of incomplete reduction. This was demonstrated by heating all three samples to 800 °C in a nonreducing atmosphere, which caused the specific magnetizations of all the samples to approach that of bulk nickel.

chemisorption—and, more generally, to study the nature of bonding at the surfaces of ferromagnetic metals.

In a study of nickel supported on SiO_2, Al_2O_3, and SiO_2–Al_2O_3, the contributions of reducibility and crystallite size to differences in specific activities for ethane hydrogenolysis were investigated by magnetic methods[29] (see Figure 2.20). The specific magnetizations of Ni in various samples with different supports and different weight loadings of Ni were used to calculate crystallite sizes. Extents of reduction were determined by sintering the Ni in nonreducing atmospheres and comparing the resultant magnetizations to those of bulk nickel. The assumption was that the sintering produced bulk nickel only from the Ni that was already reduced.

The study showed that the support strongly affected crystallite size, but weakly affected the reducibility of the nickel. Differences in ethane hydrogenolysis activity among the catalysts were not explained by measured differences in the states of the nickel; an additional effect of the support seemed necessary to explain the results.

2.10 Summary

The characterization of supported metals in catalysts is indispensable. Measurable attributes directly affect the performance of the catalyst, and many such effects are recognized and understood. Thus, characterization can direct the production of a better catalyst. Also, catalyst consistency and quality are predictable only to the extent that the catalyst is well described.

A regimen for metals characterization should include appropriate chemisorption measurements to quantify the dispersion of the metal and X-ray diffraction to verify the metal phase and provide an estimate of the volume-averaged particle size. Electron microscopy is also a reasonable tool for obtaining an initial description of the metals in the catalyst.

The other methods described above are more often applied when diagnosing unexpected behavior or catalyst failure; an understanding of the catalytic process will help in determining the causes. Consideration should also be given to the changes in the metal's morphology, oxidation state, or purity; such changes are often caused by particle growth, encapsulation, and poisoning.

References

1 R. J. Matyi, L. H. Schwartz, and J. B. Butt. *Catal. Rev., -Sci. Eng.* **29**, 41, 1987.

2 J. H. Sinfelt. *Bimetallic Catalysts: Discoveries, Concepts, and Applications.* Wiley, New York, 1983.

3 A. R. Miedema. *Z. Metallkde.* **69**, 455, 1978.

4   N. Yao, G. E. Spinnler, R. A. Kemp, and D. C. Guthrie. *Proceedings of the 49th Annual Meeting of the Electron Microscopy Society of America*. (G. W. Bailey, Ed.) San Francisco Press, San Francisco, 1991, pp. 1028–1029.

5   O. Beeck. "Catalysis and the Absorption of Hydrogen on Metal Catalysts," In *Advanced Catalysis II*. (W. G. Frankenburg, V. I. Komarewsky, and E. K. Rideal, Eds.) Academic Press, New York, 1950.

6   B. M. W. Trapnell. *Chemisorption*. Academic Press, New York, 1955.

7   S. Brunauer, P. H. Emmet, E. Teller. *J. Am. Chem. Soc.* **60**, 309, 1938.

8   J. J. F. Scholten and J. A. Konvalinka. *Trans Faraday Soc.* **65**, 2465, 1969.

9   D. J. C. Yates, L. L. Murrell, and E. B. Prestridge. *J. Catal.* **57**, 41, 1979.

10   M. Boudart and H. S. Hwang. *J. Catal.* **39**, 44, 1975.

11   G. B. McVicker, R. T. K. Baker, R. L. Garten, and E. L. Kugler. *J. Catal.* **65**, 207, 1980.

12   M. Boudart and G. Djega-Mariadassou. *Kinetics of Heterogeneous Catalytic Reactions*. Masson, Paris, 1982, pp. 165–211.

13   J. H. Sinfelt. *Catal. Rev.* **3**, 175, 1969.

14   J. H. Sinfelt, G. H. Via, and F. W. Lytle. *J. Chem. Phys.* **72**, 4832, 1980.

15   H. P. Klug and L. E. Alexander. *X-ray Diffraction Procedures for Polycrystalline and Amorphous Materials*. Wiley, New York, 1974, p. 618.

16   H. Brumberger. *Transactions Am. Crystal. Soc.* **19**, 1, 1983.

17   M. J. Yacaman. *Appl. Catal.* **13**, 1, 1984.

18   *Synchrotron Radiation Research*. (H. Winick and S. Doniach, Eds.) Plenum, New York, 1980.

19   "Synchrotron Radiation in Materials Research," In *Proceedings of the Materials Research Society, Vol. 143*. Materials Research Society, Pittsburgh, 1989.

20   S. P. Cramer, T. K. Eccles, F. W. Kutzler, K. O. Hodgson, and L. E. Mortenson, *J. Am. Chem. Soc.* **98**, 1287, 1976.

21   R. H. Herber. *The Mössbauer Effect and Its Application in Chemistry*. (R. F. Gould, Ed.) Advances in Chemistry Series, Vol. 68. American Chemical Society, Washington, D.C., 1967.

22   N. N. Greenwood and T. C. Gibb. *Mössbauer Spectroscopy*. Chapman and Hall, London, 1971.

23   "Mössbauer Effect Methodology." (I. J. Gruverman and C. W. Seidel, Eds.) In *Proceedings of the Tenth Symposium on Mössbauer Effect Methodology*. Plenum, New York, 1976.

24   W. N. Delgass and M. Boudart. *Catal. Rev.* **2**, 129, 1969.

25 T. Ishii. In "Applications of Synchrotron Radiation," *Proceedings of the CAST Symposium, June 1988, Beijing, China.* (H. Winick, et al., Eds.) Gordon and Breach, New York, 1989.

26 W. Eberhardt, P. Fayet, D. M. Cox, Z. Fu, A. Kaldor, R. Sherwood, and D. Sondericker. *Phys. Rev. Lett.* **64**, 780, 1990.

27 M. G. Mason. *Phys. Rev. B.* **27**, 748, 1983.

28 P. W. Selwood. *Adsorption and Collective Paramagnetism.* Academic Press, New York, 1962.

29 J. L. Carter and J. H. Sinfelt. *J. Phys. Chem.* **70**, 3003, 1966.

30 D. D. Beck and C. J. Carr. *J. Catal.* **110**, 285, 1988.

31 J. Robbins. Personal communication.

32 J. H. Sinfelt, Y. L. Lam, J. A. Cusumano, and A. E. Barnett. *J. Catal.* **42**, 227, 1976.

33 S. K. Behal, M. M. Disko, R. Ayer, J. Scanlon, G. Meitzner, S. C. Fung, J. H. Sinfelt, and G. H. Via. *Proceedings of the 46th Annual Meeting of the Electron Microscopy Society of America.* (G. W. Bailey, Ed.) San Francisco Press, San Francisco, 1988.

34 G. R. Wilson and W. K. Hall. *J. Catal.* **17**, 190, 1970.

35 Measurements in collaboration with R. Frahm, Hamburg Synchrotron Laboratory, Hamburg, Germany.

36 G. H. Via, K. F. Drake, Jr., G. Meitzer, F. W. Lytle, and J. H. Sinfelt. *Catal. Lett.* **5**, 25, 1990.

Bulk Metal Oxides

JAMES F. BRAZDIL

Contents

3.1   Introduction

Oxides as Catalysts

Oxides prepared in bulk form are used extensively in the chemical industry as catalysts for a wide range of processes that produce both high-volume commodity chemicals and lower-volume, higher-value specialty and fine chemicals. The effectiveness of these oxide catalysts stems from their ability to combine several solid state properties critical to the catalytic mechanism into a single-phase or multiphase oxide. This chapter will focus not only on the determination of these properties but also on their relationship to catalysis.

Bulk metal oxide catalysts are most prominently used for selective, or partial, oxidation of hydrocarbons, aromatics, and organic molecules containing heteroatoms. Partial oxidation is the oxidation of an organic molecule short of complete combustion to carbon dioxide. Examples of industrially important partial oxidation processes that use metal oxide catalysts are: propylene oxidation to acrolein and ammoxidation to acrylonitrile using bismuth-molybdenum oxide based catalysts,

$$CH_2CHCH_3 + O_2 \longrightarrow CH_2CHCHO + H_2O$$

$$CH_2CHCH_3 + \tfrac{3}{2}O_2 + NH_3 \longrightarrow CH_2CHCN + 3H_2O$$

butane oxidation to maleic anhydride using a vanadium–phosphorus oxide catalyst,

$$n\text{-}C_4H_{10} + \tfrac{7}{2}O_2 \longrightarrow C_4H_2O_3 + 4H_2O$$

acrolein oxidation to acrylic acid over complex mixed metal oxide catalysts containing molybdenum,

$$CH_2CHCHO + \tfrac{1}{2}O_2 \longrightarrow CH_2CHC(O)OH$$

methanol oxidation to formaldehyde using iron-molybdenum oxide catalysts,

$$CH_3OH + \tfrac{1}{2}O_2 \longrightarrow H_2CO + H_2O$$

ethyl benzene dehydrogenation to styrene over mixed transition metal oxide based catalysts,

$$C_6H_5CH_2CH_3 \longrightarrow C_6H_5CH=CH_2 + H_2$$

and methane oxidative coupling to higher molecular weight hydrocarbons with, for example, lithium-magnesium oxide catalysts,

$$2CH_4 + \tfrac{1}{2}O_2 \longrightarrow C_2H_6 + H_2O$$

Other commercially important chemical processes that rely on metal oxide catalysts are dehydrogenation and hydrogenation of hydrocarbons, isomerization, and olefin metathesis. Chromium aluminum oxide is widely used as a catalyst for dehydrogenation of butane to form butenes and butadiene. The latter is used for the manufacture of synthetic rubber. Copper–chromium oxide on silica or silica alumina supports is used as the olefin polymerization catalyst for the commercial production of polyethylene.

Mechanistic Features of Oxide Catalyzed Reactions

An important first step in characterizing catalysts, including oxide catalysts, is to decide what features of the material are likely to be important in defining its catalytic behavior. Key to this decision is the understanding, to the extent possible, of the surface reaction and solid state mechanisms that are operative. It is not within the scope of this book to provide a comprehensive survey of what is known about molecular and atomic level mechanisms for the myriad reactions and oxide catalyst systems that exist.[1] It is, however, possible to state a few of the general aspects of the most common mechanisms by which oxide catalysts function.

Selective oxidation reactions, including partial oxidation of olefins and alkyl aromatics, generally proceed by what is referred to as an allylic mechanism, wherein the rate-determining step is α-hydrogen abstraction by lattice oxygen of the catalyst to form a π-allyl intermediate. Oxygen from the catalyst is then inserted into the intermediate to produce the oxygen-containing product. The lattice oxygen is

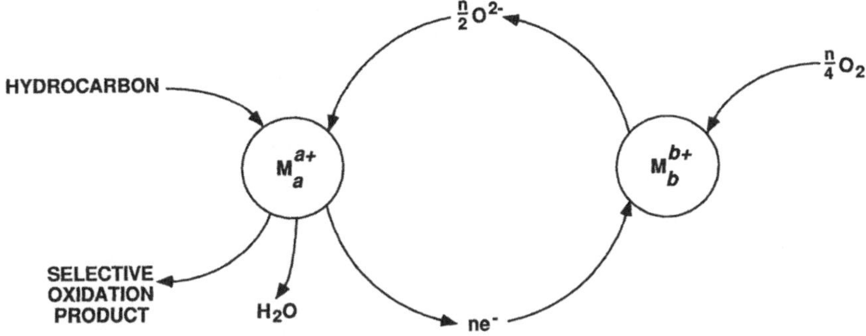

Figure 3.1   Catalytic selective oxidation–reduction cycle.

replenished by the oxygen in the bulk by a diffusion mechanism. Oxygen is ultimately replenished by reoxidation with gaseous oxygen at a surface site separate from the site of hydrocarbon oxidation. This concerted reduction–reoxidation process is termed the Mars van Krevelen mechanism. A general scheme is shown in Figure 3.1. A hydrocarbon molecule is oxidized on the surface of the oxide catalyst by the lattice oxygens of the oxide. The resulting reduced surface site is reoxidized by the transfer of electrons from this site to a separate surface site, M_b, where O_2 is reduced to O^{2-} lattice oxygens. The lattice oxygens migrate from this site through the bulk of the catalyst to reoxidize the active site M_a. For catalysts that operate by this mechanism, an understanding of the reduction and reoxidation characteristics of the oxide is important, as is information regarding the rate of oxide mobility, oxide transport mechanisms, and electronic properties that affect reducibility and electron transfer.

Oxides also effect oxidative transformations through adsorbed surface oxygen species generated by the interaction of O_2 with surface sites.

$$O_2(ads) \longrightarrow O_2^-(ads) \longrightarrow O_2^{2-}(ads) \longrightarrow 2O^-(ads) \longrightarrow 2O^{2-}(lattice)$$

Such species are generally important for the conversion of paraffins, especially the oxidative dehydrogenation of paraffins to olefins. Ethane is oxidatively dehydrogenated to ethylene over CoO/MgO with surface O^- serving to activate ethane by hydrogen radical abstraction. Clearly, for such a catalytic process, information about surface defect structure, cation coordination, and oxidation state—all of which can serve as centers for reductive chemisorption of O_2—is important for the characterization of the catalyst system.

3.2   Synthesis Methods

When characterizing oxide catalysts, it is critical to know the chemistry by which the catalyst was prepared. It is well-appreciated by those working in the catalysis field that the synthesis method can profoundly affect catalyst surface structure and composition, morphology, and even bulk solid state phase composition. Figuratively speaking, the catalyst is said to remember how it was prepared even after being subjected to numerous heat treatments and activation procedures at high temperatures. Therefore, to correctly interpret data collected in the course of characterizing an oxide catalyst, it is usually valuable to know the preparative chemistry that gave rise to its physical and structural properties.

Oxide catalysts used in industrial processes are invariably multi-metal oxides. Single metal oxides are normally prepared as model catalysts for fundamental research investigations. Practical oxide catalysts may be single phase, but they are usually multiphase. Also, they are usually in the form of powders (10 to 100 μm), pellers, or tablers that are polycrystalline or amorphous. Both unsupported catalytically active oxides and oxides supported on high surface area materials such as silica, alumina, or refractory monoliths are encountered in commercial operation.

Mixed metal oxide catalysts are usually prepared by coprecipitation, thermal transformation of a precursor compound, sol–gel formation, or grinding and firing a metal oxide mixture. In its broadest definition, coprecipitation is the process by which a solution containing soluble compounds of the metal components for the catalyst is treated—usually by adding an acid or base to change pH—to effect the formation of a precipitate from the solution. The resulting precipitate is typically a simple mixture of the hydrous oxides of the metals, although more complex chemistry and compound formation may occur, as will be discussed later. The precipitate is transformed into the final mixed metal oxide catalyst by removing the water and heat-treating the dried material to decompose the hydrous oxides and salts. High-temperature treatment is then used to effect the solid state reactions that produce the intimately mixed oxide phases, compounds, or solid solutions, or any combination of these, that are the active species of the catalyst.

Catalyst preparation from specially prepared precursors is, in many ways, a special case of coprecipitation. The method is particularly effective for preparing mixed metal oxide systems in which the catalytically active phase is a specific multi-metal oxide compound or mixed metal solid solution. Precursors may be formed by precipitating a compound from solution and decomposing or transforming the precursor by treatment at high temperature, sometimes in a controlled-atmosphere environment. An example is the preparation of zinc copper hydroxycarbonate, $(Cu,Zn)_2(OH)_2CO_3$, by the combination of a solution of copper nitrate and zinc nitrate with a solution of $NaHCO_3$, followed by washing the resulting carbonate precipitate to remove sodium. The mixed metal hydroxycarbonate is then heat-treated in order to drive off CO_2 and produce a copper zinc spinel.[2] In other cases the precursor is more than just a heat decomposable template for one or more metals. It provides the structural framework necessary for the controlled crystal growth of the active oxide phase. This is the case for vanadium–phosphorus oxide catalysts for butane oxidation to maleic anhydride. The catalytically active $(VO)_2P_2O_7$ phase forms from the topotactic conversion of $VOHPO_4 \cdot \tfrac{1}{2}H_2O$ that

         BULK METAL OXIDES   Chapter 3

is generated from the reaction of V_2O_5 and H_3PO_4 in a nonaqueous medium such as isobutanol.[3]

$$2VOHPO_4 \cdot \tfrac{1}{2}H_2O \longrightarrow (VO)_2P_2O_7 + 2H_2O$$

The sol–gel method of catalyst preparation is also related to the coprecipitation method. In general, a colloidal dispersion containing the metal constituents of the final catalyst is first formed and then made to gel by changing pH, cation concentration, or temperature. The resulting gel is heat-treated to form an intimate mixture of individual metal oxides, compounds, or solid solutions.[4]

Grinding and heating mixtures of metal oxides is not widely used in the preparation of practical catalysts. The method generally produces materials with low surface areas, and thus low catalytic activity, due to the high temperature normally required for the requisite solid state reactions to occur between metal oxides. However, the method is used to produce highly crystalline materials that serve as models of complex catalyst systems. These materials are amenable to detailed structural analyses, such as X-ray diffraction, for determining crystal structures and solid state phase composition.

The application of oxides to the surface of a carrier can be accomplished by impregnation by using a solution of the soluble compounds of the metal components of the catalyst. Spraying or coating a carrier with a dispersion of insoluble compounds in water or nonaqueous medium is also commonly used. The resulting material is subjected to high temperature treatment to produce the desired oxide compounds on the surface of the carrier.

3.3 Properties of Oxides and Their Relation to Catalytic Behavior

Determination of Bulk Structure

The primary objective of determining the bulk structure of metal oxide catalysts is the correlation of catalytic behavior and catalytic phenomena with specific solid state phases. By this means, attempts can be made to identify the key components and solid state structural features of a catalyst that are responsible for causing chemical transformations; these features are associated with specific aspects of the molecular mechanism of the catalyzed reaction. Armed with these insights, the catalyst scientist can develop rational strategies for improving the efficiency of a catalyst by tailoring modifications around the key constituents. Numerous studies have been published that develop empirical correlations between catalytic and solid state structural aspects of a range of oxide catalysts.[5] These "structure–function relationships" constitute a valuable data base from which one can rationalize observed catalytic behavior and put in perspective new data on catalytic phenomena.

Among the most powerful techniques for bulk structure determination are X-ray, neutron, and electron diffraction methods. The X-ray diffraction method is the most widely used, primarily because of easy sample preparation and the availability of convenient laboratory X-ray sources. Experimental approaches to bulk structure determination[7] have been described in the lead volume in this series, *Encyclopedia of Materials Characterization*. Powder diffraction techniques are most commonly used by catalyst scientists, since most of the catalysts used in laboratory studies and in commercial operation are prepared in the form of polycrystalline powders. Single-crystal materials are less commonly used for characterization studies, but they can provide important information about catalytic mechanisms. They generally are prepared as models of working catalysts for carefully studying specific catalytic phenomena and catalyst features through the use of sensitive surface science techniques[6] or for determining an unknown crystal structure for a component of the catalyst.

The varied crystal structures that metal oxides adopt have been described in detail by Wells[7] and others and will not be reviewed here. Instead, this chapter concentrates on the interpretation of such information about the bulk structure of oxides for understanding catalyst behavior.

The first information generally to be gleaned from a bulk structure by X-ray or neutron diffraction is the identification of the crystalline solid state phases that are present. This is accomplished by comparing the diffraction pattern with the known patterns for compounds which can possibly be present based on the constituents of the catalysts. It is possible, even for a complex mixed metal oxide system containing copper, zinc, cobalt, and chromium, to completely identify the phases present by assigning all the diffraction lines to the possible oxides that can form with these metals.[2]

When only one solid state phase is identified in a mixed metal oxide system, it is likely that a solid solution has formed between the constituent oxides. Evidence for solid solution formation is obtained by indexing the X-ray diffraction pattern, calculating the unit cell parameters for the single phase material (see the *Encyclopedia of Materials Characterization*), and then examining the changes in these cell parameters as function of catalyst composition. A monotonic, usually linear, change in unit cell size commensurate with the relative size of the substituting metal is good evidence that a random solid solution exists (see Figure 3.2). It is generally necessary to prepare and examine a series of model compositions in order to identify the most probable substitution scheme for solid solution formation.[8] This approach can also be used to probe the location in a structure of a metal that is present in more than a single oxidation state in the compound, as in the the case of lead and bismuth ruthenate catalysts for electrochemical oxygen reduction.[9] Lead and bismuth have been found to substitute in specific crystallographic sites in the structure based on the relative size of the cations present: Pb^{2+}, Pb^{4+}, Ru^{4+}, Ru^{5+}, Bi^{3+}, and Bi^{5+}.

Solid solution formation is a powerful catalyst design strategy that provides a means to tailor catalytic behavior by integrating the catalytic properties of the individual metal oxides. Thus, it is possible to have these individual catalytic properties work in concert to effect a complex, multi-step chemical transformation

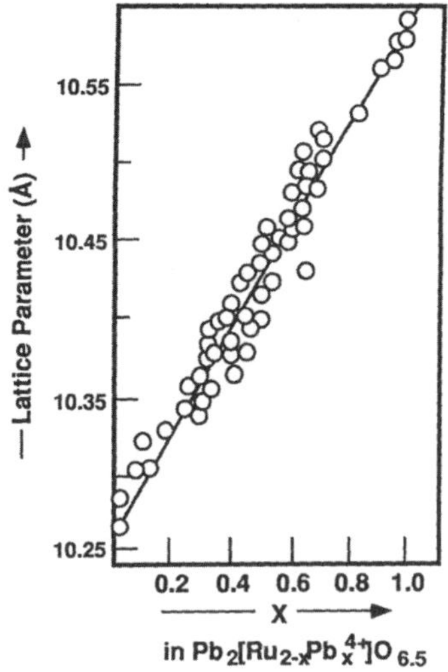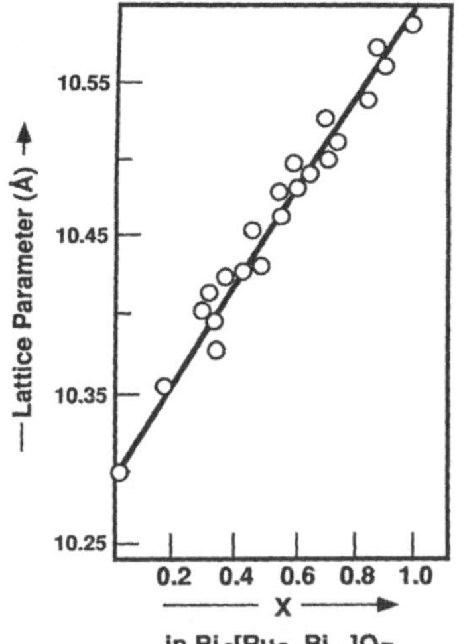

Figure 3.2 Lattice parameter versus composition. (Reproduced with permission from *Preparation and Characterization of Materials*, J. M. Honig and C. N. R. Rao, Eds., Academic Press, New York, 1981, pp. 38, 44.)

on the catalyst surface. Lead molybdate, $PbMoO_4$, which is inactive as a hydrocarbon oxidation catalyst, becomes an effective oxidation and ammoxidation catalyst by incorporation of bismuth in the form of a solid solution, $Pb_{1-3x}Bi_{2x}[]_xMoO_4$, where [] is a cation vacancy defect[10] (see Figure 3.3). Bismuth is a key constituent of the olefin oxidation catalyst responsible for activation via α-hydrogen abstraction.[11]

Bulk structural characterization of solid solutions provides a means for probing the catalytic properties of the individual metal components as well as discerning the role of other solid state structural features such as vacancy defects. In particular, a combination of bulk characterization techniques is useful for identifying catalytically important structural features. In the case of a lead molybdate based solid solution, the substitution of altervalent cations bismuth, sodium, and lanthanum for lead can be used to control the concentration of cation vacancy defects, as illustrated by the following substitution formula:[12]

$$Pb_{0.84-3x}Bi_{0.08}Na_{0.08}La_{2x}[]_xMoO_4$$

Evidence for solid solution formation is again obtained from X-ray diffraction, which reveals the presence of a single crystalline phase with changes in unit cell parameters consistent with the relative size of the substituting cations. The concentration of cation vacancy defects is inferred from the fact that the system is single phase for $x = 0$ to 0.04 and has the scheelite crystal structure, as determined by X-ray diffraction. Electroneutrality requirements based on the fixed oxidation states of the cations dictate the necessary oxygen and cation stoichiometry, which in turn

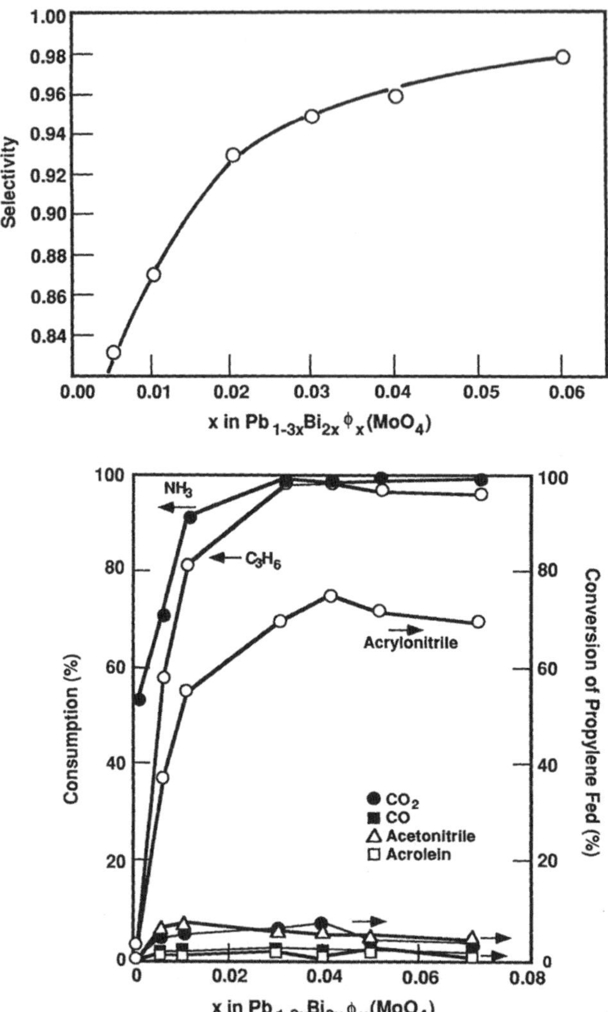

Figure 3.3   (Top) Selectivity of 1-butene oxidation as a function of cation vacancy concentration in the $Pb_{1-3x}Bi_{2x}[]_xMoO_4$ system. (Bottom) Percentage consumption and conversion during propylene ammoxidation as a function of x in the $Pb_{1-3x}Bi_{2x}[]_xMoO_4$ system. Reaction temperature is 440 °C, and feed composition is 4.0% C_3H_6, 4.8% NH_3, 47.7% air, and 43.5% N_2 with a 6.0 s contact time. (Reproduced with permission from Reference 10.)

requires cation sites in the structure to be vacant, with the vacancies distributed randomly throughout the structure. The existence of cation or oxygen vacancy defects can be corroborated by measuring crystal density using the flotation of crystals in mixed liquids having different densities. Neutron diffraction can also be used to infer the presence of cation and oxygen vacancies through its ability to precisely position cations and oxygen in a crystal structure using either a single crystal

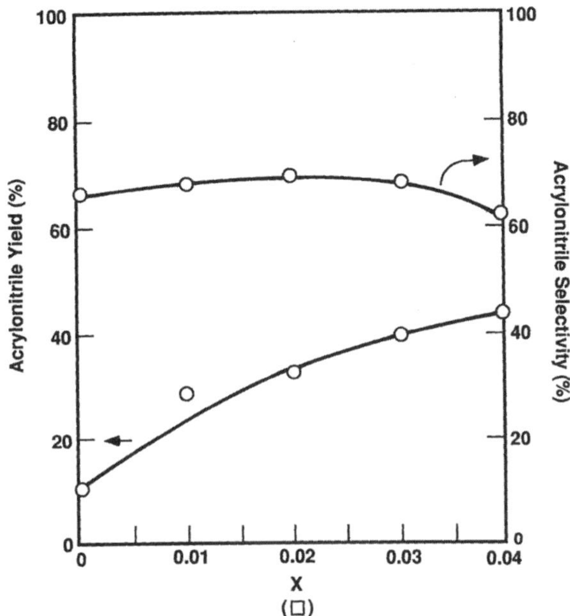

Figure 3.4 Acrylonitrile yield and acrylonitrile selectivity for propylene ammoxidation for the $Pb_{0.84-3x}Bi_{0.08}Na_{0.08}La_{2x}[]_xMoO_4$ system as a function of cation vacancy concentration. (Reproduced with permission from Reference 12.)

or polycrystalline sample. As shown in Figure 3.4, catalytic activity was found to be directly proportional to the concentration of these defects. Examination of these materials by Raman spectroscopy[12] (see Figure 3.5) showed the formation of a weak high frequency vibrational band at about 900 cm^{-1} as a shoulder on the more intense 869 cm^{-1} band. This new band was assigned to the stretching vibration of a molybdenum-oxygen double bond, a structural feature not present in $PbMoO_4$. The criticality of the Mo=O moiety to catalytic activity indicates that it plays a key role in the catalytic mechanism, most likely as active lattice oxygen for selective oxidation of propylene to acrolein or acrylonitrile.

Similarly, solid solution formation is used to enhance desirable catalytic properties by incorporating new metal components which benefit, or promote, one or more of the steps in the catalytic cycle. Structural incorporation of cerium into a bismuth molybdenum oxide catalyst, $Bi_{2-x}Ce_x(MoO_4)_3$, significantly enhances catalytic activity for propylene ammoxidation to acrylonitrile, as confirmed by both X-ray and neutron diffraction.[13] This enhanced activity is presumably due to the multivalent character of cerium, which is consequently expected to promote surface and lattice oxygen mobility, an important feature of selective oxidation catalysts.[14]

A more common occurrence in characterizing the bulk structure of oxide catalysts is the identification of more than a single crystallographically distinct phase. To understand the catalytic behavior of such a system, it is necessary to first

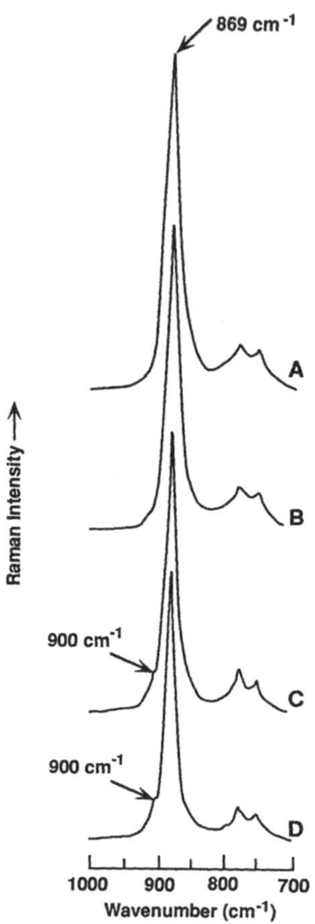

Figure 3.5 $Pb_{0.84+x}Bi_{0.08}Na_{0.08-2x}MoO_4$ system Raman spectra: (A) $x = 0$; (B) $x = 0.02$; (C) $x = 0.03$; and (D) $x = 0.04$. (Reproduced with permission from Reference 12.)

determine the importance, if any, of each phase to the overall performance of the catalyst by correlating changes in catalyst composition—and thus changes in phase composition—to one or more catalytic properties, such as selectivity for formation of a product, degree of conversion of a reactant, and long-term performance characteristics. It generally is not a trivial task to unambiguously identify all the solid state phases that may be present. In many cases, this requires systematic preparation of binary, ternary, and quaternary oxides of varying composition in order to confidently identify all phases existing in a single catalyst system. Furthermore, the possibility exists that phases may be present in the catalyst that are amorphous or microcrystalline, and thus these will not be detectable by diffraction methods. In this case, Raman and infrared vibrational spectroscopies can be helpful in identifying non-crystalline phases. The potential complexity is illustrated by the seemingly simple cerium–tellurium–molybdenum oxide catalyst system. Sorting through the rich phase chemistry of this three metal oxide catalyst reveals that the most active and selective catalyst composition contains as many as five distinct phases, with the catalytically most important being α-$Ce_2Mo_4O_{15}$ and $Ce_4Mo_{11}Te_{10}O_{59}$.[15]

   BULK METAL OXIDES   Chapter 3

In many cases, the catalytic behavior of a multiphase oxide cannot be simply attributed to the sum of the catalytic properties of the individual phases that comprise the catalyst system. Synergistic effects may occur which challenge the catalyst scientist's ingenuity in identifying the basis for the interactions between the phases. This is generally done by correlating specific catalytic phenomena with systematic variations in the relative phase composition and phase purity, as discerned from X-ray diffraction and electron microscopy.

Bulk Composition

The determination of the metal component composition of an oxide is a seemingly routine, yet quite important, part of characterizing an oxide catalyst. There are two reasons for carefully determining bulk composition. One is for correlating metal component composition with the surface composition to ascertain if there is any preferential enrichment or depletion of catalyst components on the surface. This makes it is possible to ascribe particular catalytic behavior to a specific metal or constituent oxide phase present on the active surface of the catalyst.

The second reason is quality assurance. In the case of preparative procedures that involve filtration, solid–liquid separations, or coating, or when a constituent oxide of a mixed oxide system is volatile under the heat treatment conditions used to prepare the catalyst, it is clearly important to determine how the final composition compares with the intended composition of the catalyst.

The preferred methods for bulk composition determination are elemental analyses by the atomic absorption (AA) and atomic emission–inductively coupled plasma (ICP) methods. These methods require dissolution of the oxide in acid—usually in nitric acid, hydrochloric acid, or hydrofluoric acid. The methods provide a sensitivity of at least 10 ppm, although this is dependent upon the element and possible spectral interference from other elements that are present. Another option is X-ray fluorescence (XRF). Compared to ICP, XRF requires little sample preparation, since the catalyst powder can be analyzed without the need for dissolution of the sample. The XRF method is most sensitive to the heavier elements. The sensitivity and accuracy of the XRF method can be compromised by so-called matrix effects, in which partial absorption of the fluorescence emission by the oxide may occur.

Metal Oxidation State and Local Structural Environment

The importance of determining the oxidation states of the metal constituents in oxide catalysts becomes clearly evident from the wide range of catalytic behavior that a metal oxide exhibits, depending on its valency and the valence distribution in the solid. In order to understand the functioning of a mixed metal oxide catalyst containing multivalent metals, it is critical to characterize the valency distribution of the metal constituents. It is most desirable, if possible, to obtain this information while under the catalyst's typical operating temperature, pressure, and gas compositions.

Closely related to metal oxidation state determination is information about the local structural environment of the metals in the oxide lattice. Changes in metal oxidation state are usually accompanied by changes in the local structure around the metal. Valency and local structure are also closely related with respect to measurement techniques: analytical methods to determine one will also provide information, or at least allow for inferences, about the other.

Metal oxidation state is frequently determined using solution methods by first dissolving the metal constituents of the oxide in a strong, non-oxidizing acid such as sulfuric, hydrochloric, or hydrofluoric and then analyzing by redox titration or through electrochemical means such as polarography. Redox titrations have been applied to vanadium oxide containing catalysts[16] with potassium permanganate as the redox titrant. Such solution methods, though simple from the standpoint of instrumentation, have several drawbacks. Complete solubility of the catalyst sample is not always possible. Also, care must be taken to insure that the cation in the solid is not altered upon dissolution; this frequently necessitates deoxygenating the acid, which itself is no guarantee that a redox reaction has not occurred when several multivalent metals are present. A further disadvantage is that titration techniques are not element specific: solution methods generally provide only the average or total oxidation state and not the distribution of oxidation states of a metal or between several metals in complex oxides. Finally, such methods can never be in situ analyses of oxidation state under the typical high-temperature conditions in which catalysts operate.

For paramagnetic elements, electron spin resonance (ESR) has proven useful for obtaining element-specific information about oxidation states. It is also possible to extract information about the structural environment of the element from the data. The method has most frequently been applied to molybdenum and vanadium oxide catalysts, since V^{4+} and Mo^{5+} give strong spin resonance signals. However, because of the possibility of spin coupling due to cation pairing, as in the case of V^{4+} in the $(VO)_2P_2O_7$ catalyst, it generally is not possible to use ESR to quantify oxidation state in a solid. At best, ESR can determine the presence of a spin active cation, but it cannot provide unequivocal evidence of its absence.

Similarly, solid state NMR using magic angle spinning (MAS-NMR) can be a sensitive, element-specific probe of the local environment of a cation in an oxide lattice provided there are no paramagnetic cations or species in the solid. For example, ^{51}V solid state NMR has been used to determine the coordination state of vanadium in a complex bismuth vanadium oxide catalyst.[17] The local chemical environment and site symmetry can be obtained by comparing the observed chemical shifts and line shapes with model compounds that are structurally well-characterized.

Mössbauer spectroscopy has proven to be a valuable method for obtaining information about the distribution of oxidation states for Mössbauer-active elements. The vast majority of studies have been conducted on catalysts containing iron, tin, or antimony because of the relative availability of gamma ray sources for these

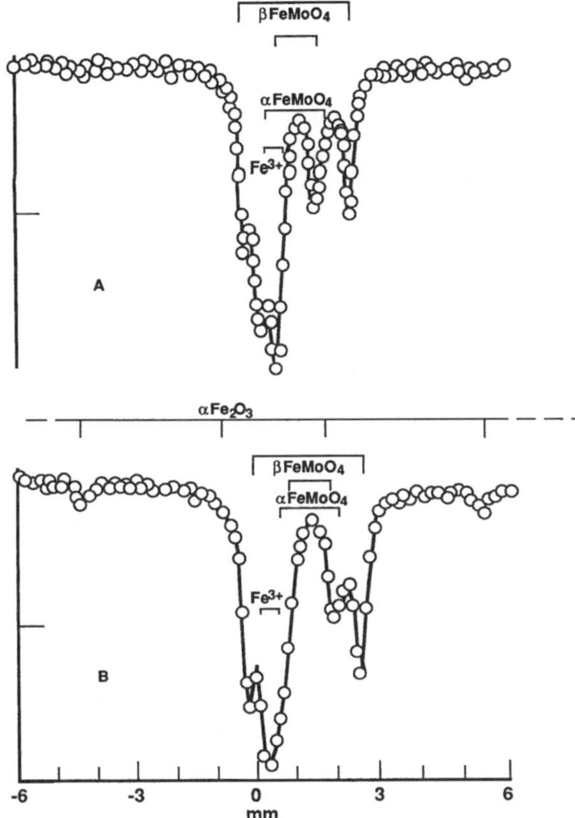

Figure 3.6   Mössbauer spectra of iron cobalt molybdate catalyst recorded at (A) 295 K and (B) 4 K. (Reproduced with permission from Reference 19.)

elements. Mössbauer spectra contain element-specific information on local symmetry, oxidation state, and structural environment, including the extent of cluster formation and phase composition, as shown in work on iron containing ZSM-5 zeolite catalysts[18] and iron molybdate based oxidation catalysts[19] (see Figure 3.6). Such information can be obtained for iron, tin, and antimony, in which oxidation state and site symmetry are revealed by characteristic isomer shifts and quadrupole splittings.

An extremely powerful technique for oxidation state and metal coordination determination is X-ray absorption near-edge structure (XANES) spectroscopy. As one component of extended X-ray absorption fine-structure (EXAFS) analysis, XANES provides information about local structure in a solid from spectral features that are easily correlated to oxidation state by comparison with well-characterized model compounds.[20] XANES has several important advantages over the techniques already described in this chapter. These include the elimination of the need to alter the physical state of the catalyst sample, the element specificity of the method, and the ability to examine the catalyst under reaction conditions (since it is an X-ray based spectroscopy). The use of XANES for catalyst characterization is limited, however, since it requires use of a synchrotron radiation source. Some of the most successful applications of XANES to oxide catalyst studies have been with vanadium oxide based systems. In these cases, XANES was able to reconcile the conflicting findings from other analyses and unambiguously identify vanadium oxidation state and vanadium–oxygen coordination.[21, 22]

Characterization by Electron Microscopy

The oxide catalyst characterization techniques discussed so far in this chapter all provide chemical and structural information of the composite oxide. It is only with electron microscopy (and most recently with scanning tunneling microscopy, or STM, which is still being developed as a characterization technique) that nearly atomic-level resolution analysis can be achieved on solid materials. Scanning transmission electron microscopy (STEM), transmission electron microscopy (TEM), and high resolution electron microscopy (HREM) have been applied extensively to the characterization of a wide range of oxide catalysts.[23] It is possible to obtain structural and chemical analysis down to as little as 2 Å using specialized HREM techniques and sample preparation (using high electron beam voltages up to 500 kV and very thin specimens in the range of 10 to 100 Å). Resolution between 1 and 20 nm, readily achieved with conventional, commercially available STEM and TEM, is sufficient to address the most commonly encountered chemical and structural questions related to oxide catalyst performance.

Electron microscopy (EM) is not inherently capable of in situ catalyst characterization, since it requires high vacuum conditions for electron detection and imaging. However, progress has recently been made in microscope design and in electron detection equipment that may eventually allow routine EM analysis with reactant gases over the catalyst sample.

Electron microscopy provides an important additional window on catalytic behavior of oxides that the bulk characterization techniques cannot. For example, EM can identify low concentration or impurity phases in a catalyst by imaging across a 1 μm (or less) catalyst particle. The distribution of solid state phases, identified by X-ray diffraction, can be determined in this way. An inhomogeneous distribution of phases in a multiphase oxide catalyst can be directly responsible for less than optimum catalyst performance. Elemental analysis, using the characteristic X-rays produced by a heavy element of the oxide in the electron beam, can also provide this information. The location, distribution, and concentration of the metal constituents of the oxide can be determined, making it possible to identify, and thus to understand, the functioning of either a poisoning or a promoting element in the catalyst. Elements added to a catalyst that affect catalytic behavior generally will associate preferentially with specific phases of the catalyst. EM can readily discern the location of these elements and thereby help in assigning roles for the phases that are present in the catalyst.

Low resolution analysis of a catalyst sample to determine crystal morphology and observe crystal habits of phases in the oxide also provides useful information about an oxide catalyst. Changes in crystallite size and crystal orientation in the catalyst can influence catalytic behavior, since such differences result in changes in surface area and the surface concentration of catalytically important metal constituents of the oxide. It is becoming increasingly evident that the different crystal planes of a solid oxide have profoundly different catalytic behavior,[24, 25] which is due to differences in metal concentration or metal–oxygen coordination at the different crystal faces. Thus, it is possible in many instances to understand the basis for a catalyst's performance by examining crystal orientation through EM. With this information it can be possible to develop variations in catalyst synthesis or thermal treatment that affect the orientation of the key catalytic phases of the oxide so as to obtain optimum catalytic performance.

HREM, being a more specialized and experimentally more demanding technique than STEM and TEM, is generally used when specific questions arise regarding structural defects, the orientation of the phases with respect to each other (e.g., epitaxy and coherence at phase boundaries), and the possibility of formation of localized structures, microdomains, and structural intergrowths that can play important roles in the functioning of oxide catalysts. For example, the intergrowth of three bismuth tungstate phases on an atomic scale shows up clearly by HREM imaging, as shown in Figure 3.7. HREM has proven valuable in oxide catalyst characterization (with the discovery of new structure types that are nonequilibrium or are stabilized by coherent intergrowth within a host phase), in elucidation of the mechanism of solid state transformation during thermal or catalytic treatments, and for the identification of order and disorder defect structures with near atomic-level resolution.[26] An excellent example of this is the identification of the localized oxygen vacancy defect structure produced in a calcium manganese oxide selective oxidation catalyst by the removal of lattice oxygens in the course of the catalytic cycle. Computer simulation combined with HREM imaging provides evidence for oriented microdomains of the nonstoichiometric perovskite phase $CaMnO_{2.8}$, having specific ordering of oxygen vacancies in the (001) layers.[27]

Surface Characterization Using Molecular Probes

Although surface characterization will be discussed extensively in chapters on supported catalysts, there are several techniques for surface characterization that are important for understanding bulk oxide catalysts. One method that is ubiquitous in both bulk and supported catalyst characterization is nitrogen surface area determination.[28] The method has become universally accepted for measuring the amount of catalyst surface available for interaction with the reactant molecules. This amount should be viewed as an upper limit value, since in nearly all cases the

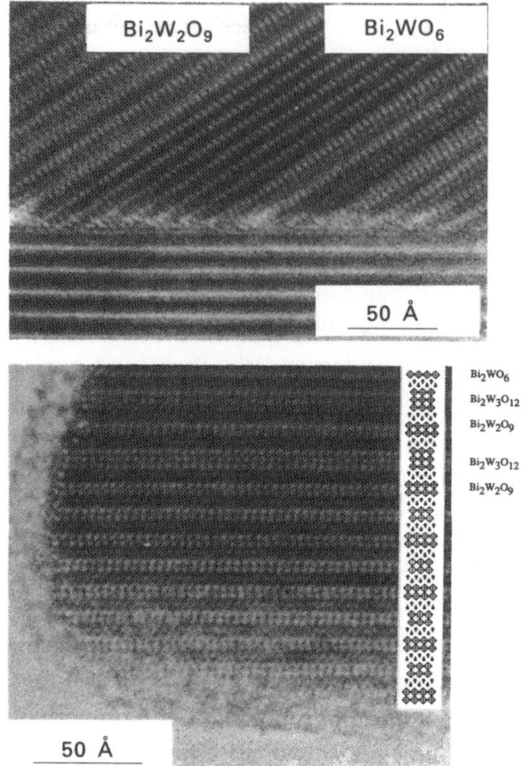

Figure 3.7 (Top) HREM image of intergrowths in the bismuth tungsten oxide system. (Bottom) HREM image of bismuth tungsten oxide intergrowths with computer simulation showing metal–oxygen coordination corresponding to each intergrowth: $Bi_2W_3O_{12}$, $Bi_2W_2O_9$, Bi_2WO_6.

catalytically active surface is only a small fraction of this total surface area. Nevertheless, surface area measurements are valuable for gauging the activity, or potential activity, of the oxide catalyst. Surface area is frequently an important physical feature of the catalyst that must be optimized in order to obtain the best catalyst performance. Optimization of surface area is generally done by modifying the time or temperature or both for heat treatment or by changing one or more variables in the preparation chemistry of the oxide.

Another common characterization method used with oxide catalysts is pore volume determination using the mercury intrusion technique.[29] As in the case of nitrogen surface area, the method is used to judge the interactive availability of the reactive surface of a catalyst by assessing—based on pore size, pore distribution, and the molecular dimensions of the reactant and product molecules—the contribution of the inner pore structure of the catalyst to catalytic activity.

Another method for surface characterization uses strongly interacting probe molecules to obtain a measure of the actual reactive surface of a bulk oxide or to infer the number of active surface sites. The most frequently used technique is acid–base titration, in which, for example, a basic molecule such as ammonia or pyridine is made to adsorb on the surface of an oxide catalyst. The amount adsorbed is readily measured by the pressure decrease over the catalyst or by the weight gain by the catalyst. This amount is taken, in this case, as the number of surface acid sites, assuming one base molecule adsorbs on one acidic site. Similarly, acidic molecules, such as carboxylic acids, are used to probe basic sites. The method is frequently used in conjunction with infrared spectroscopy in order to determine the nature of the structure of the adsorbed molecule and thereby determine the type of acid or base site—Brönsted or Lewis. A drawback to this characterization method is that it is generally done at a temperature well below that at which the oxide is active as a catalyst, since it is necessary to prevent conversion or decomposition of the probe molecule before its concentration on the surface can be determined. Thus, it is problematic how this adsorption information can be related to the nature of the surface of the oxide catalyst under high temperature reaction conditions.

Other chemisorptive probes can be used to characterize the surface of oxides. Oxygen has frequently been used to examine both fully oxidized and partially reduced oxide surfaces. These studies generally are directed toward ascertaining the concentration and nature of those surface oxygen species that can play a role in catalysis.

Surface Characterization Using Spectroscopic Analyses

Spectroscopic analyses of the surfaces of bulk oxides are used by catalysis scientists primarily to determine elemental composition and surface oxidation states. The most frequently used techniques are X-ray photoelectron spectroscopy (XPS), ultraviolet photoelectron spectroscopy (UPS), Auger electron spectroscopy (AES), and secondary-ion mass spectroscopy (SIMS). These are all described in detail in the *Encyclopedia of Materials Characterization*. Since all these spectroscopic techniques require ultra-high vacuum conditions, in situ characterization of the catalyst surface is not feasible. These surface analysis instruments, however, can be equipped with external cells for high temperature and pressure treatment of the catalyst followed by evacuation and insertion into the vacuum chamber.

The determination of elemental composition at the surface of a multi-metal mixed oxide catalyst using these methods is done quite easily and the results are usually unambiguous with respect to the relative ratios of the elements present in the surface region. Frequently, the information provides key insights into catalytic behavior. A comparison of the surface composition with that of the bulk shows immediately if a particular metal constituent or oxide phase is preferentially enriched at the catalyst surface and is thereby dominating the catalysis. It is not uncommon to find that an element that has been added to the bulk as a promoter at a low level is enriched on the surface by a factor of 10 to 100 relative to the other elements present. It is critical to know such surface characteristics before drawing any conclusion about the performance of a complex catalyst system as a function of its bulk composition.

In contrast, the correlation of oxidation states measured by XPS or UPS with catalytic behavior is more problematic. Many of the multivalent elements, even when stabilized in the bulk of an oxide in a low oxidation state, tend to be present on the surface at or near their highest oxidation state. This is largely due to the tendency of such surface elements to oxidize when exposed to an oxygen environment or to oxygen contamination during sample preparation or transfer. Even with treatment in a controlled-environment cell or through sputtering to remove a surface oxidized layer, it is not possible to ensure that the measured surface oxidation states reflect those present in the working catalyst.

Reducibility and Oxide Ion Mobility

As previously noted in this chapter (see Section 3.1), oxides that are effective as catalysts for partial oxidation reactions generally function by a redox mechanism involving cycles of lattice oxygen loss (reduction) and reoxidation of the reduced oxide by O_2. Thus, the effectiveness of oxides to serve as selective oxidation catalysts can be largely determined by their propensity for oxidation state cycling and the facile movement of oxide ions between the bulk and surface of the oxide.

Several approaches have been used to measure the redox proclivity of metal oxide catalysts. The methods which have shown good correlations with catalytic behavior have been those that effectively decouple the reduction and reoxidation in the catalytic cycle at or near the typical operating temperature and pressure conditions of the catalyst. One approach uses a pulse technique.[30] This method entails passing sequential pulses of a reducing gas—usually hydrocarbon, CO, NH_3, or H_2—over the catalyst at an elevated temperature in the absence of co-fed oxygen. From the analysis of the resulting oxidized products, it is possible to determine the degree and rate of lattice oxygen depletion from the catalyst. In turn, reoxidation is probed by sequential pulsing of the resulting reduced catalyst with an O_2-containing gas to determine the amount and rate of oxygen uptake by the catalyst. The most informative results are generally obtained when the reducing pulses contain the same reactant or reactants as are used in the catalytic process. The conclusions from the experiments can then be more confidently applied to understanding the redox process that the catalyst undergoes under steady state catalytic conditions.

Other effective approaches to understanding lattice oxygen participation are temperature-programmed reduction and temperature-programmed reoxidation.[31] Reduction involves contacting the catalyst with the reactant gas in the absence of oxygen at a temperature below the reaction temperature for oxidation, then increasing the temperature linearly. Conversely, reoxidation is performed by contacting the reduced catalyst with oxygen at a similarly low temperature and again increasing the temperature linearly.

The kinetic results obtained from pulse or temperature-programmed methods are generally the rates of lattice oxygen removal and re-incorporation into the oxide as well as activation energies for the reduction and reoxidation processes.[30, 31] Interpretation of the kinetic data is less straightforward than other methods. The mechanism of reduction will depend on the reducing gas used, since a specific chemical interaction between the gas molecule and the catalyst surface is a requisite first step. This interaction will depend on the chemical nature of the molecule and specific features of the oxide, including surface composition and structure, grain boundary structure, structural defects (such as shear planes) that can provide migration paths for lattice oxygen between the surface and the bulk, and phase changes that produce suboxides. Particular care must be taken to ascertain if structural changes and phase segregation occur as a result of the reduction step. This is generally done by X-ray diffraction analysis of the catalyst after reduction.

One of the most useful pieces of information that can be gleaned from a controlled catalyst reduction experiment is the extent of lattice oxygen participation in the catalysis. This information is readily obtained from the pulse experiment described above by comparing product distributions from the reaction in the presence and in the absence of co-fed oxygen. The same product distribution for both indicates that oxygen from the oxide is serving as the oxidizing agent under catalytic conditions and that O_2 is serving to reoxidize these centers in a mechanistically distinct step. A divergent distribution suggests that co-fed oxygen is playing a more direct role in the catalysis, probably through the formation of a surface-activated oxygen species. In this case, other techniques are needed to understand the nature of these oxide species.

Magnetic and Electronic Properties

The measurement of magnetic properties and the determination of electronic structure have generally proven ineffective for understanding catalytic properties of bulk metal oxides. Some exceptions to this exist, particularly when the measurements are viewed alongside data from other techniques as a means to fully characterize the solid state properties of an oxide. Most notable has been the use of magnetic susceptibility measurements, at times in combination with Mössbauer spectroscopy, to determine the size of iron oxide particles on a support.[18, 32] Of course, this technique is limited to oxides that exhibit magnetic properties. The oxides most frequently studied using magnetic measurements are iron, nickel, cobalt, and manganese oxides.

Direct information about the electronic structure of an element in an oxide lattice is obtained using XPS. As described earlier in this chapter, XPS has proven to be most useful as a tool for determining surface elemental composition and surface metal oxidation state. Electronic structure determination is most useful for understanding photocatalytic, photoassisted surface reactions and electrochemical processes on semiconductor oxides such as TiO_2. In these instances, the surface chemistry is dictated by electronic transitions across the band gap of the oxide, from the valence to the conduction bands. A surface reaction is initiated by the transfer of a conduction band electron to a reactant species at the surface. Knowledge of the electronic structure of the oxide, including band gap energy and electronic defect states produced by bulk elemental doping, can be used to modify the catalyst in order to change the energetics of the electronic transitions and thereby affect the rate, and potentially the mechanism, of the surface reactions.

The role of electronic transitions in thermal catalyzed processes on semiconductor oxides has been debated extensively. The "electronic theory of catalysis" was used at one time in an attempt to explain a wide range of oxide catalyzed reactions.[33] It is generally recognized today that a bulk electronic structure view of catalytic phenomena is inadequate to explain selectivity in catalysis, and the theory has found limited use as a strategy for the design of catalyst systems. Thus, electronic structure determination of oxide catalysts is generally used to augment information about other characteristics of an oxide that play a more direct role in the catalysis, such as metal coordination and oxidation state.

3.4 Summary

Just as no single characterization method can be used to fully explain the basis for the catalytic behavior of an oxide catalyst, haphazard application of multiple characterization techniques will not ensure useful insights into catalytic phenomena. It is still (and will continue to be) left to the judgment of the catalysis scientist, especially as new analytical instrumental techniques continue to become available for solid state characterization, to select the battery of techniques[34] that will give information that is most germane for the development of improved catalysts and new catalytic processes.

Among the most critical properties one needs to know to characterize bulk oxide catalysts are:

- bulk crystalline structure

- bulk and surface elemental composition

- oxidation states of the metal constituents

- local structural environment of the metal constituents

- concentration and disposition of structural defects, such as cation and anion vacancies

- surface area and bulk porosity

- redox behavior and oxide reducibility.

Acknowledgments

The author wishes to thank L. C. Glaeser, M. A. Antonio and A. W. Varnes for helpful discussions during the preparation of this chapter. The author also acknowledges the help and guidance provided by colleagues over the years on characterizing oxide catalysts, particularly R. G. Teller, D. D. Suresh, G. L. Shoemaker, and E. Kostiner.

References

1   A. Bielanski and J. Haber. *Catal. Rev.-Sci. Eng.* **19**, 1, 1979.

2   M. Piemontese, F. Trifiro', A. Vaccari, E. Foresti, and M. Gazzano. In "Preparation of Catalysts V," *Proceedings Fifth Int. Symp. Scientific Bases for Preparation of Heterogeneous Catalysts, Studies in Surface Science and Catalysis, Vol. 63.* (G. Poncelet, P. A. Jacobs, P. Grange, and B. Delmon, Eds.) Elsevier, Amsterdam, 1991, p. 49.

3   V. A. Zazhigalov, G. A. Komashko, A. I. Pyatnitskaya, V. M. Belousov, J. Stoch, and J. Haber. In "Preparation of Catalysts V," See Ref. 2, p. 497.

4   C. E. Marsden. In "Preparation of Catalysts V," See Ref. 2, p. 215.

5   *Solid State Chemistry in Catalysis.* (R. K. Grasselli and J. F. Brazdil, Eds.) American Chemical Society, Washington D.C., 1985.

6   G. A. Somorjai. *Chemistry in Two Dimensions: Surfaces.* Cornell University Press, Ithaca, NY, 1981.

7   A. F. Wells. *Structural Inorganic Chemistry*, 5th ed. Clarendon, Oxford, 1984.

8   J. Ziolkowski, R. Kozlowski, K. Mocala, and J. Haber. *J. Solid State Chem.* **35**, 297, 1980.

9   H. S. Horowitz, J. M. Longo, H. H. Horowitz, and J. T. Lewandowski. In *Solid State Chemistry in Catalysis.* See Ref. 5, p. 143.

10   A. W. Sleight. *Advanced Materials in Catalysis.* (J. J. Burton and R. L. Garten, Eds.) Academic Press, New York, 1977, p. 181.

11   R. K. Grasselli, J. D. Burrington, and J. F. Brazdil. *Farad. Discuss. Chem. Soc.* **72**, 203, 1982.

12   J. F. Brazdil, L. C. Glaeser, and R. K. Grasselli. *J. Catal.* **81**, 142, 1983.

13   J. F. Brazdil, R. G. Teller, R. K. Grasselli, and E. Kostiner. In *Solid State Chemistry in Catalysis.* See Ref. 5, p. 57.

14   G. W. Keulks and L. D. Krenzke. *Proc. Int. Congr. Catal., 6th, 1976.* (G. C. Bond, P. B. Wells, and F. C. Tompkins, Eds.) The Chemical Society, London, 1977, p. 806.

15   J. C. J. Bart, N. Giordano, and P. Forzatti. In *Solid State Chemistry in Catalysis.* See Ref. 5, p. 89.

16 G. Centi, D. Pinelli, F. Trifiro', C. Fumagalli, L. Capitanio, and G. Stefani. *Chim. Ind.* Milan. **72**, 625, 1990.

17 F. D. Hardcastle, I. E. Wachs, H. Eckert, and D. A. Jefferson. *J. Solid State Chem.* **90**, 194, 1991.

18 L. N. Mulay and T. Pannaparayil. In *Catalyst Characterization Science—Surface and Solid State Chemistry.* (M. L. Deviney and J. L. Gland, Eds.) American Chemical Society, Washington D.C., 1985, p. 498.

19 B. Benaichouba, P. Bussiere, G. Coudurier, H. Ponceblanc, and J. C. Vedrine. *Hyperfine Interactions.* **57**, 1741, 1990.

20 J. Wong, F. W. Lytle, R. P. Messmer, and D. H. Maylotte. *Phys. Rev. B.* **30**, 5596, 1984.

21 S. Asbrink, G. N. Greaves, P. D. Hatton, and K. Garg. *J. Appl. Cryst.* **19**, 331, 1986.

22 T. Tanaka, H. Yamashita, R. Tsuchitani, T. Funabiki, and S. Yoshida. *J. Chem Soc. Faraday Trans. 1.* **84**, 2987, 1988.

23 C. E. Lyman. In *Catalyst Characterization Science—Surface and Solid State Chemistry.* See Ref. 18, p. 361.

24 J. C. Vedrine, G. Coudurier, M. Forissier, and J. C. Volta. *Catal. Today.* **1**, 261, 1987.

25 A. Andersson and S. Hansen. *J. Catal.* **114**, 332, 1988.

26 J. M. Thomas. *Ultramicroscopy.* **8**, 13, 1982.

27 A. Reller, D. A. Jefferson, J. M. Thomas, and M. K. Uppal. *J. Phys. Chem.* **87**, 913, 1983.

28 "Standard Test Method for Surface Area of Catalysts," *Annual Book of ASTM Standards.* D3663-84. American Society for Testing and Materials, Philadelphia, 1988.

29 "Standard Test Method for Determining Pore Volume Distribution of Catalysts by Mercury Intrusion Porosimetry," *Annual Book of ASTM Standards.* D4284-83. American Society for Testing and Materials, Philadelphia, 1988.

30 J. F. Brazdil, D. D. Suresh, and R. K. Grasselli. *J. Catal.* **66**, 347, 1980.

31 T Uda, T. T. Lin, and G. W. Keulks. *J. Catal.* **62**, 26, 1980.

32 J. Phillips, Y. Chen, and J. A. Dumesic. In *Catalyst Characterization Science—Surface and Solid State Chemistry.* See Ref. 18, p. 518.

33 F. F. Vol'kenshtein. *The Electronic Theory of Catalysis on Semiconductors.* Macmillan, New York, 1963.

34 J. F. Brazdil, M. Mehicic, L. C. Glaeser, M. A. S. Hazle, and R. K. Grasselli. In *Catalyst Characterization Science—Surface and Solid State Chemistry.* See Ref. 18, p. 26.

Supported Metal Oxides

ISRAEL E. WACHS AND KOHICHI SEGAWA

Contents

4.1 Introduction

Supported metal oxide catalysts consist of a metal oxide component deposited on the surface of a second metal oxide substrate that usually has a high surface area. The deposited metal oxide component is the active phase of the catalyst. Typical active oxides are those of rhenium, chromium, molybdenum, tungsten, vanadium, and niobium; and typical high–surface area materials that are employed as substrates or supports are alumina, silica, titania, niobia, and zirconia.

Supported metal oxide catalysts find wide application in the petroleum, chemical, and pollution-control industries.[1] Alumina-supported rhenium oxide catalysts are commercially employed for olefin metathesis reactions and petroleum refining. Silica-supported chromium oxide catalysts are used as olefin polymerization catalysts, and alumina-supported chromium oxide catalysts are used for the dehydrogenation of butane to butenes and in early reforming processes. Alumina-supported molybdenum oxide catalysts find application in the petroleum industry as hydro-desulfurization (HDS) and hydrodenitrogenation (HDN) catalysts. Alumina-supported tungsten oxides are used as catalysts in hydrodesulfurization, hydrodenitrogenation, hydrogenation, and hydrocarbon cracking. Titania-supported tungsten oxide is introduced as a promoter for commercial DeNOx catalysts that selectively reduce NOx with ammonia from stationary emissions. Supported vanadium oxide is used extensively as an oxidation catalyst for the partial oxidation of hydrocarbons, the ammoxidation of hydrocarbons, and the selective catalytic reduction of nitrogen oxides from stationary emissions. Supported niobium oxide catalysts catalyze partial oxidation and hydrocarbon cracking reactions. Supported titanium oxide catalysts promote the olefin polymerization reaction over silica-supported chromium oxide and are capable of epoxidizing olefins. The catalysis literature contains numerous other applications of supported metal oxide catalysts.

4.2 Synthesis Methods

Supported metal oxide catalysts are synthesized by depositing the active metal oxide on the surface of an oxide support (i.e., alumina, silica, titania, etc.). Many different methods have been developed for accomplishing this.[2] The active component can be introduced as a precursor or salt in an appropriate solvent, aqueous or nonaqueous, by impregnating the solution using the incipient wetness method. During the incipient wetness preparation method, the solvent and the metal oxide precursor are first absorbed into the pores of the oxide support through capillary action. The solvent is subsequently removed by drying, and the precursor is converted to the metal oxide by calcination in air at elevated temperatures. The nature of the solvent, aqueous or nonaqueous, is dictated by the properties of the specific metal oxide precursor (solubility in the solvent, stability in aqueous environments, and stability in air). Typical metal oxide precursors are ammonium salts (aqueous), oxalates (aqueous), and alkoxides (nonaqueous and air sensitive). The simplicity of this preparation procedure makes the incipient wetness preparation the most common synthesis method.

An equilibrium adsorption preparation method has recently been developed that controls the amount of active metal oxide deposited on the oxide support by controlling the pH of the aqueous solution[3]; this method is used mainly in fundamental studies. The relationship between the molybdenum oxide loading and the pH of the impregnating solution is shown in Figure 4.1.

The active metal oxide can also be deposited on the surface of the support from the gas phase or liquid phase by employing a suitable volatile precursor (metal chloride or oxychloride).[4] This synthesis method is referred to as grafting since the metal oxide precursors directly react and titrate the surface hydroxyls of the support.

Also, in the last few years it has been discovered that metal oxides can also be deposited on the surface of oxide supports by physically mixing and heating the active component and the support material (thermal spreading).[5] However, this method applies only to active metal oxides that are volatile or have a low melting temperature, such as rhenium oxide, molybdenum oxide, tungsten oxide, and

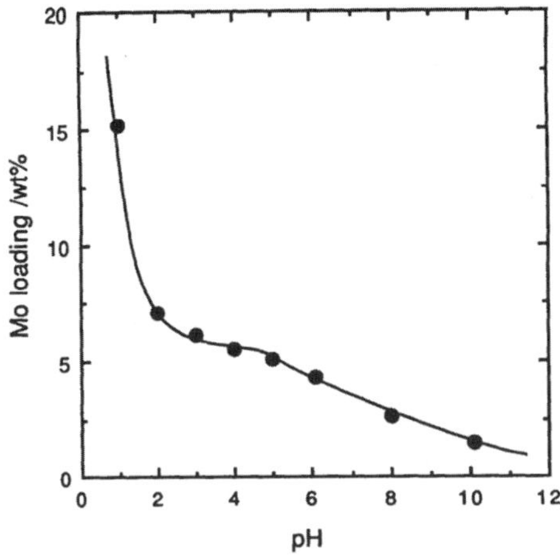

Figure 4.1   Relation between Mo loadings and pH of the impregnating solution.

vanadium oxide. The disadvantage of this method is the long calcination times required to achieve complete spreading of the active metal oxide over the support surface.

Recent characterization studies[6] of supported metal oxide catalysts prepared by various synthesis methods reveal that all preparation methods result in essentially the same supported metal oxide catalyst. Thus, the specific preparation method for supported metal oxide catalysts is not as critical as was once thought.[2]

4.3 Characterization

Structure of the Supported Metal Oxide Phase

Deposition of an active metal oxide onto an oxide support can result in either a two-dimensional surface metal oxide overlayer or in three-dimensional metal oxide crystallites. Large, three-dimensional crystallites can be detected by using X-ray diffraction (XRD), but crystallites below approximately 40 Å and two-dimensional metal oxide overlayers cannot be detected by XRD. Thus, the absence of an XRD peak may be due to small crystallites, a low concentration of crystallites, or to the presence of a two-dimensional metal oxide overlayer. The presence of an XRD peak, however, confirms only that large crystallites are present and does not provide information about the presence or absence of a two-dimensional metal oxide overlayer.

Fortunately, this can be determined by using Raman spectroscopy, which is able to discriminate among these various possibilities. Unlike XRD, the Raman signal is not dependent on long-range order and depends only on the vibrations of the M-O bonds. The different metal oxide structures that are present in the two-dimensional metal oxide overlayer and in the oxide crystallites produce different Raman signals, which readily allow discrimination between these two structures. This is shown in the in situ Raman spectra in Figure 4.2 for a series of oxidized MoO_3/TiO_2 catalysts, in which the 998 cm^{-1} band results from the

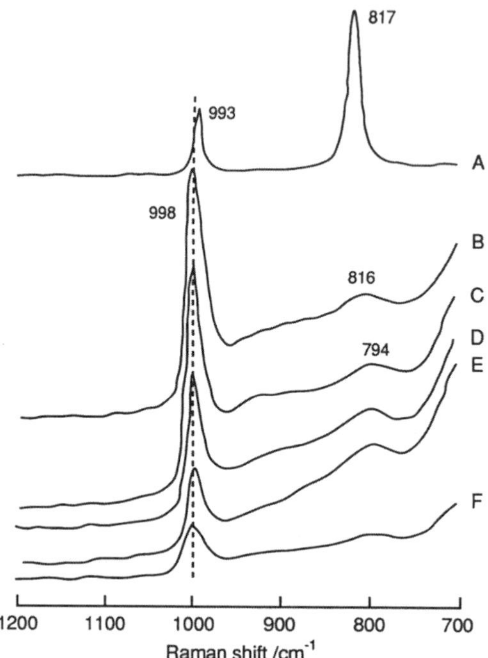

Figure 4.2 Laser Raman spectra of MoO₃/TiO₂ catalysts as a function of pH of the impregnating solution (taken after being calcined at 773 K in the in situ cell): (A) 20.0 wt %, pH = 2.6; (B) 7.1 wt %, pH = 2.6; (C) 6.6 wt %, pH = 4.0; (D) 4.2 wt %, pH = 7.3; (E) 3.5 wt %, pH = 8.7; and (F) 2.4 wt %, pH = 9.7.

two-dimensional molybdenum oxide overlayer and the 817 cm^{-1} band results from MoO$_3$ crystallites.[6] Infrared spectroscopy (IR) can also sometimes be used to discriminate between the metal oxide structures, but the strong absorption of the IR signal by the oxide support in the region of interest usually does not make this a straightforward approach.[7] Extended X-ray absorption fine-structure spectroscopy (EXAFS—discussed in detail in Chapter 2) can also discriminate between the metal oxide structures, but this technique averages the signal from all sites, making it very difficult to discriminate among several structures simultaneously present in the catalyst. Furthermore, the additional difficulty associated with performing EXAFS experiments at specialized synchrotron radiation sources makes this method the least effective technique for determining monolayer coverage. The different reduction characteristics of the two-dimensional metal oxide overlayer and the crystallites usually allows discrimination between these two metal oxide phases through temperature-programmed reduction (TPR) experiments, which result in different reduction peaks for each phase.[8] The TPR method is, however, an indirect method, and Raman spectroscopy is the method of choice for determining the structure, two-dimensional or three-dimensional, of the supported metal oxide phase.

     SUPPORTED METAL OXIDES   Chapter 4

Surface Coverage of the Supported Metal Oxide Phase

In situations where a two-dimensional metal oxide overlayer is present, it is important to determine the extent of surface coverage of the support by the supported metal oxide phase. The maximum amount of two-dimensional active metal oxide that can be accommodated on the support is referred to as monolayer coverage. Additional active metal oxide beyond the monolayer coverage usually results in the formation of metal oxide crystallites because the support cannot accommodate any additional metal oxide in the two-dimensional overlayer. Thus, Raman spectroscopy can directly determine the monolayer coverage because it detects the additional bands due to the crystalline metal oxide phase. This was already demonstrated by the in situ Raman spectra in Figure 4.2 for the titania-supported molybdena system, in which MoO_3 crystallites are formed above monolayer coverage. The crystalline-phase Raman bands are usually significantly stronger than those of the two-dimensional metal oxide overlayer for many metal oxide systems, so even trace amounts of crystallites are detectable by Raman spectroscopy. In Figure 4.2, the lowest molybdenum oxide loading, in which MoO_3 crystallites—corresponding to Raman band at 817 cm^{-1}—are detected on the titania support, corresponds to 7.1% MoO_3 (sample prepared at a pH of 2.6). The Raman band at 794 cm^{-1} that is present at lower molybdenum oxide loadings originates from the titania support.[6] Consequently, monolayer coverage of molybdenum oxide on this titania support, which has a surface area of 55 m^2/g, occurs slightly below 7.1% MoO_3, since this loading already exceeds monolayer coverage.

The two-dimensional metal oxide overlayer is stabilized on the support by reaction with the surface hydroxyls; this allows IR to determine the monolayer coverage, since all the surface hydroxyls are consumed at monolayer coverage.[9] Similarly, carbon dioxide titrates the basic surface hydroxyls of supports such as alumina and titania by forming surface carbonates, and at monolayer coverage essentially no surface hydroxyls remain on the support surface.[10] Thus, carbon dioxide chemisorption can also determine the monolayer coverage. This is demonstrated for the MoO_3/TiO_2 system in Figure 4.3, wherein the carbon dioxide chemisorption is monitored as a function of the pH of preparation (the relationship between preparation pH and MoO_3 content is given by Figure 4.1) as well as the BET surface areas. The carbon dioxide uptake becomes zero at a preparation pH of approximately 4, which corresponds to approximately 6.6% MoO_3 and suggests that monolayer coverage is achieved in this region.

The different X-ray photoelectron spectroscopy (XPS) cross-sections of the two-dimensional metal oxide overlayer and metal oxide crystallites also may be used for the determination of monolayer coverage. In these studies, a dramatic change in slope occurs when the XPS surface signal is plotted as a function of bulk composition because of these very different cross-sections. The XPS cross-section for the two-dimensional phase is significantly greater than that for the agglomerated crystalline phase, since not all the metal oxide can be detected by the XPS measurement

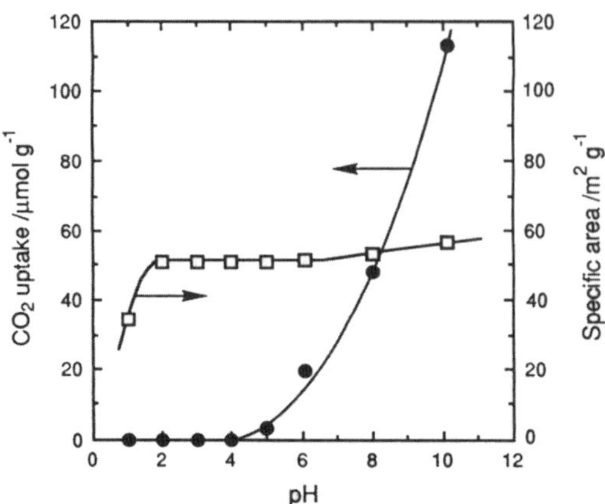

Figure 4.3   BET surface areas and CO_2 chemisorption of MoO_3/TiO_2 as a function of pH values. Samples were measured after calcination at 773 K.

in the crystalline phase due to the limited depth analysis of this technique. Figure 4.4 shows the XPS analysis of the MoO_3/TiO_2 system as a function of molybdenum oxide loading. At low molybdenum oxide loading, the surface Mo signal, which is represented by the XPS Mo(3*d*)/Ti(2*p*) ratio referenced against the titania support signal, increases linearly with the bulk Mo content, which is represented by the MoO_3 loading, because every Mo atom in this two-dimensional structure is detected by the XPS measurement. However, at loadings above 6.6% MoO_3, which corresponds to a pH of 4.0, the surface Mo signal deviates from linearity with increasing bulk Mo content because of the formation of three-dimensional MoO_3 crystals. Thus, in agreement with the above Raman and carbon dioxide characterization studies, the XPS measurements also reveal that the surface molybdenum oxide monolayer coverage is achieved at approximately 6.6% MoO_3 loading on this titania support.

Similar experiments can also be conducted using ion scattering spectroscopy (ISS) as the characterization method. However, the ISS technique is not as commonly used nor as easy as XPS. Temperature programmed reduction (TPR) experiments can also determine the monolayer coverage by means of the different reduction temperatures of the two-dimensional metal oxide overlayer and the metal oxide crystallites.[8] Many comparative measurements (Raman, IR of surface hydroxyls, carbon dioxide chemisorption, XPS, ISS, and TPR) have demonstrated that the same monolayer coverages of the two-dimensional metal oxide overlayers are obtained independent of the method of measurement.

In some circumstances (e.g., metal oxides supported on a weakly interacting support like silica), crystalline phases can be formed before complete monolayer coverage is achieved; multiple characterization methods may be required to establish such behavior. Such cases are revealed by using both Raman spectroscopy to

                                  SUPPORTED METAL OXIDES   Chapter 4

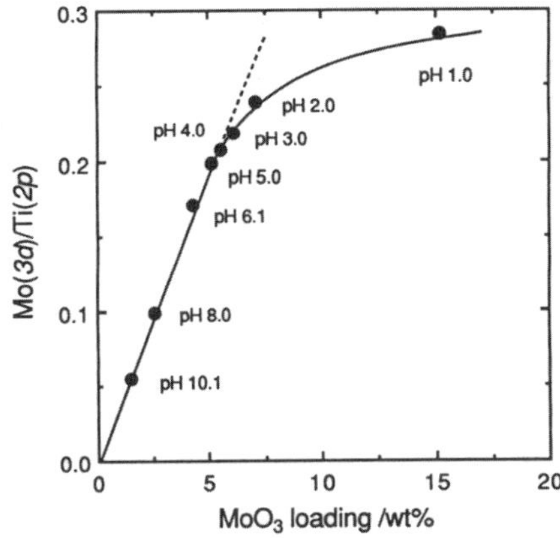

Figure 4.4 XPS Mo(3d)/Ti(2p) ratio as a function of Mo loadings.

detect the crystallites and carbon dioxide chemisorption or IR to obtain information about the consumption of surface hydroxyls; but it is usually sufficient to use Raman spectroscopy alone since there tends to be a weak interaction between the deposited metal oxide and the oxide support, which makes the crystallites appear at significantly lower metal oxide loadings than would be expected with only monolayer coverage. For example, silica-supported vanadium oxide reveals the presence of V_2O_5 crystallites at 1 to 2% vanadia loading, whereas approximately 10% vanadia loading would be required to account for every 100 m^2/g of the silica support.[8]

Oxidation States and Local Structural Environments of Supported Metal Oxide Phases

The oxidation states of the two-dimensional metal oxide overlayers can be directly determined by the chemical shifts present in the XPS measurements.[11] Such studies have shown that the metal oxide overlayers are always in their maximum oxidation states after calcination treatments. These fully oxidized overlayers are also detected with Raman and IR spectroscopy, since they usually give rise to metal-oxo (M=O) vibrations in the 900–1050 cm^{-1} region; upon reduction, these features are absent. The oxidation states can also be measured with X-ray absorption near-edge spectroscopy (XANES) experiments, since the location of the pre-edge features is directly related to the oxidation state of the metal oxide.[13] However, XPS is the most commonly used method to determine oxidation states because of its availability, ease of measurement, and straightforward interpretation.

Much effort has been directed in the past few years to determine the molecular structures of the amorphous and disordered fully oxidized two-dimensional metal oxide overlayers present in supported metal oxide catalysts. Raman spectroscopy gives rise to a signal directly related to the local structural environment of the metal oxide as well as to the M–O bond length.[12, 14] Thus, Raman spectroscopy has received the most attention for use in determining the molecular structures of the two-dimensional metal oxide overlayers. IR's inability to observe many of the metal-oxygen vibrations because of the strong absorption of the supports has resulted in its limited use for making structural determinations. However, the complementary nature of IR and Raman has assisted some of the Raman assignments.[7, 14] It is possible to obtain additional detailed structural information on those metal oxide systems that possess nuclei that give a strong solid state nuclear magnetic resonance (NMR) signal. The solid state NMR measurements for supported vanadium oxide catalysts have significantly contributed to the understanding of the molecular structures of these catalytic systems.[15] Solid state NMR measurements for supported phosphorous oxide[16] and silica[17] have also been very successful. Unfortunately, corresponding solid state NMR studies with supported tungsten oxide and niobium oxide do not provide solid state NMR signals[18]; and at present, molybdenum oxide only provides signals that are not sensitive to the structure of the surface metal oxide species and are sometimes difficult to interpret.[19] However, more recent solid state Mo NMR studies appear very promising.[17] Additional structural information about the two-dimensional metal oxide overlayers can also be provided by X-ray absorption measurements (XANES and EXAFS).[20–22] The pre-edge features present in XANES provide information about the coordination of the metal oxide (four-fold or six-fold), and EXAFS contains information about the M–O distances. The one drawback of XANES and EXAFS is that these techniques are not molecular spectroscopies (such as Raman, IR, and NMR), so they average over all sites if more than one structure is present (a situation that is quite common with supported metal oxide catalysts). However, after other characterization techniques reveal that only one site is present, the XANES/EXAFS information is very useful.

In contrast to the significant progress made in recent years on the oxidation states and local structural environments of fully oxidized supported metal oxide phases, very little is known about reduced supported metal oxide phases. Reduction studies, such as TPR and thermogravimetric analysis (TGA) provide information about the average oxidation state of reduced oxide phases. Direct information about the oxidation states present upon reduction can be obtained from XPS measurements. However, several different oxidation states may coexist on the surface of a reduced sample, which complicates the spectral analysis and may make curve fitting necessary.[11] Raman and IR spectroscopy may not be as useful for the reduced states because the absence of terminal M=O bonds results in much weaker signals; also, the vibrational assignments are currently not understood. Solid state NMR studies will not be useful for reduced samples since the presence of paramagnetic species such as V(+4), Cr(+3), Cr(+5) and Mo(+5) interferes with the NMR measurement by producing severe signal broadening. In contrast, electron paramagnetic resonance (EPR) is an excellent technique because it is sensitive to paramagnetic species. However, EPR can only detect specific reduced oxidation states. In situ EXAFS/XANES have not been performed on such reduced supported metal oxide systems, and the possibility of several different oxidation states existing raises many doubts about this nonmolecular spectroscopy method. In summary, detailed information about the oxidation states and, especially, the molecular structures of reduced two-dimensional metal oxide overlayers is very difficult to obtain at present, the two most informative techniques being XPS (for the oxidation state information) and EPR (for the structures of paramagnetic states).

Morphology of the Supported Metal Oxide Phase

Several attempts have been made to obtain micrographs using high-resolution transmission electron microscopy (HR-TEM) that show the morphology of the two-dimensional metal oxide overlayers in supported metal oxide catalysts. However, the two-dimensional metal oxide phases are very difficult to observe because of the lack of contrast between the supported oxide phase and the underlying oxide support and because of the disordered state of these overlayers compared to crystalline oxides.[23] A further complication is that the surface metal oxide overlayer appears to agglomerate under the high intensity of the electron beam typically employed in such experiments.[23–25] (In contrast, three-dimensional crystalline metal oxide phases in supported metal oxide catalysts are readily observed in such experiments and are not altered by the electron beam.[23]) However, studies with model silica supports are proving to be successful, and the surface metal oxide phases can be observed under ambient conditions.[23, 26] Under ambient conditions, surface metal oxide phases are present as clusters (see discussion below on ambient structures), which may facilitate their detection; no measurements have yet been performed under in situ conditions in which these clusters decompose into smaller units. It may require several more years to determine how much morphological information can be obtained about these two-dimensional overlayers by using HR-TEM studies.

Surface Chemistry of Supported Metal Oxides

Much information about the surface chemistry of supported metal oxide catalysts can be obtained by the use of appropriate probe molecules. The adsorption of basic molecules, such as pyridine or ammonia, provides information about surface acidity.[27] Information about Lewis acid sites and Brönsted acid sites, as well as surface concentrations and strengths, can be obtained from pyridine or ammonia adsorption experiments. IR can discriminate between the protonated adsorbates associated with Brönsted acid sites and the unprotonated adsorbates associated with Lewis acid sites. The strengths of the surface acid sites are determined by examining the influence of temperature on the adsorbate surface concentrations.

The reduction characteristics of the surface metal oxide phases during TPR experiments, which typically employ hydrogen as the reducing gas, are very important. The reducibility of the surface metal oxide phases have recently been shown to be related to the catalytic activity of the supported metal oxides during partial

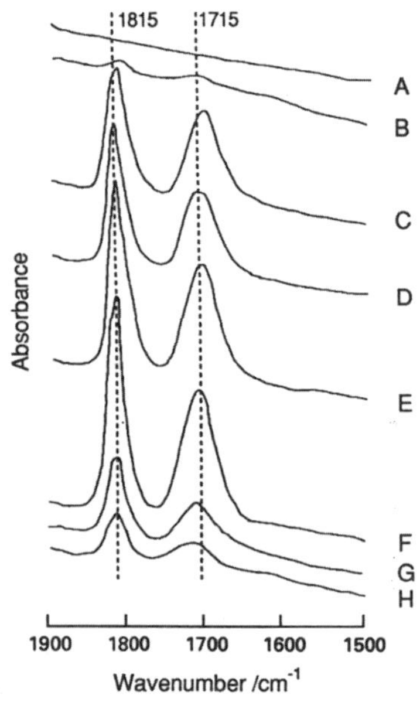

Figure 4.5 IR spectra of NO chemisorbed on MoO_3/TiO_2 (6.6 wt %, pH = 4.0) as a function of the extent of reduction: (A) $MoO_{3.0}$, (B) $MoO_{2.8}$, (C) $MoO_{2.3}$, (D) $MoO_{1.9}$, (E) $MoO_{1.7}$, (F) $MoO_{1.5}$, (G) $MoO_{1.3}$, (H) $MoO_{0.6}$.

oxidation reactions.[28] For example, the catalytic reaction rates for partial oxidation reactions are directly related to the degree of ease in removing oxygen from the surface metal oxide phase by the reactant molecules. Thus, it is very important to quantify the reducibility of the surface metal oxide phases and the influence of different catalyst parameters (surface coverage, promoters, nature of oxide support, etc.) on reducibility.

Information about the presence of reduced surface sites in the supported metal oxide phase can be obtained by NO chemisorption studies, since NO does not adsorb on oxidized sites and on the oxide support. Such experiments are typically monitored using IR. Figure 4.5 shows the IR spectra of NO chemisorbed on MoO_3/TiO_2 catalyst as a function of extent of reduction for a catalyst possessing a complete molybdenum oxide monolayer. NO chemisorbs as dinitrosyl or dimers on such surfaces, and the IR bands at 1815 and 1715 cm^{-1} are due to the associated symmetric and asymmetric stretching modes, respectively. Additional information about the adsorbed NO molecules is provided by temperature-programmed desorption (TPD) experiments which can reveal the number of different adsorption sites present on the reduced surface. This is presented in the NO TPD spectra of Figure 4.6 for the reduced MoO_3/TiO_2 catalyst as a function of extent of reduction. Clearly, there appear to be two different reduced Mo sites for NO adsorption in this catalyst, and their relative ratio varies as a function of extent of reduction. Information about the reduced surface sites is critical for catalytic reactions that occur only on reduced surface metal oxide phases (i.e., metathesis, hydrogenation, polymerization, etc.).

   SUPPORTED METAL OXIDES   Chapter 4

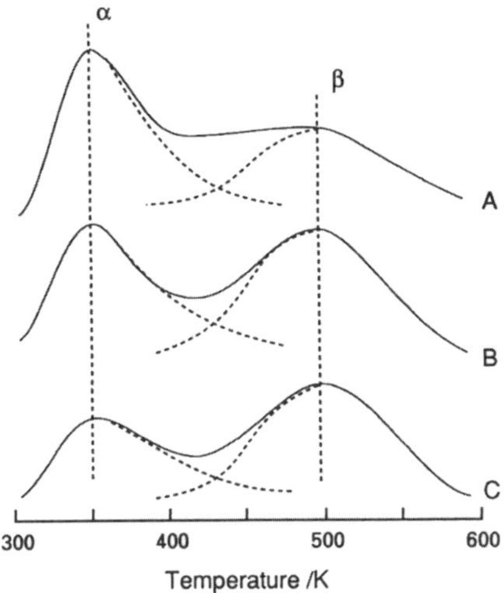

Figure 4.6 TPD spectra of NO adsorbed on MoO_3/TiO_2 (6.6 wt %, pH = 4.0) at different extents of reduction: (A) $MoO_{1.3}$, (B) $MoO_{1.8}$, and (C) $MoO_{2.3}$. The heating rate is 2.5 K/min.

Characterization Summary

The characterization of supported metal oxide catalysts requires the use of several techniques because no one technique can provide all the information required. Raman spectroscopy is the most important structural characterization method for the fully oxidized surface metal oxide phase because it performs in situ measurements and it has the ability to discriminate among the various metal oxide structures. Infrared spectroscopy is the most important technique for revealing surface chemistry, since it provides in situ information about the nature of the adsorbed species, surface acidity, and the oxide support surface hydroxyls. Information about the reactivity of the surface metal oxide phase can be obtained through temperature-programmed reduction studies, but the most sensitive probe of reactivity is the actual catalytic reaction. For partially reduced phases, X-ray photoelectron spectroscopy is best suited for providing information about the distribution of oxidation states, and electron paramagnetic resonance gives the best structural information about specific paramagnetic species.

4.4 Impregnating Solution Chemistry

At one time it was believed that it was possible to determine the final structure of supported metal oxide catalysts by controlling the aqueous chemistry of the impregnating solutions used in their preparation.[3] We now know that this is not possible, since the final catalyst structure has no memory of its preparation history due to rapid surface diffusion and the equilibration of the surface metal oxide phases.[6] Nevertheless, it is instructive to examine the solution chemistry of metal oxides, since the aqueous solution chemistry is related to the metal oxide structures present under ambient conditions (as will be discussed below) wherein the catalyst surfaces are hydrated by moisture.[29]

The aqueous solution chemistries of many metal oxide systems have been examined by Raman spectroscopy, which is able to penetrate through water because water is a weak Raman scatterer.[30] In the case of molybdenum oxide aqueous solutions, the reported Raman studies are very thorough; the following equilibria were found among the various species:

$$7MoO_4^{2-} + 8H^+ \rightleftharpoons Mo_7O_{24}^{6-} + 4H_2O$$

$$Mo_7O_{24}^{6-} + 3H^+ + HMoO_4^- \rightleftharpoons Mo_8O_{26}^{4-} + 2H_2O$$

The isolated Mo species possess tetrahedral coordination and are present in basic solutions (with high values of pH). The polymerized species are octahedrally coordinated and are present in acidic solutions (with low values of pH). This aqueous chemistry can also be monitored with Mo-95 NMR, as shown in Figure 4.7, in which the spectra of 0.1 M molybdenum solutions at different pH values of impregnating solution were obtained at 17.4 MHz. The chemical shifts were referenced to a 2 M aqueous solution of Na_2MoO_4. In basic solutions (pH $\geq$ 7.61), only the sharp resonance line of four coordinated Mo species (MoO_4^{2-}) is observed at -1.75 ppm. In acidic solutions (pH $\leq$ 5.26), a broad resonance peak is observed at 33.3–25.8 ppm, which is attributed to six coordinated Mo species containing MoO_6 groups ($Mo_7O_{24}^{6-}$ and $Mo_8O_{26}^{4-}$). The resonance shift from 33.3 to 25.8 ppm in the acidic solution is attributed to the transition from the Mo_7 to the Mo_8 species. At intermediate values of pH (plots C through E in Figure 4.7), both four and six coordinated Mo species are observed in the spectra, and the relative intensities of each peak vary as a function of pH. The relationship between the aqueous solution chemistry and the surface metal oxide molecular structures present under ambient conditions is described below.

4.5   Supported Metal Oxide Catalysts Under Ambient Conditions

Under ambient conditions, the surface of the oxide support is hydrated, with many monolayers equivalent of water, and the surface metal oxide overlayer is essentially in an aqueous medium. The amount of water present on the oxide support surface can be directly determined by TGA; it usually has been found to correspond to several weight per cent of the catalyst and is a strong function of the support BET surface area.[15] The net pH of this thin film of water is controlled by the point of zero charge (PZC), which is the surface pH at zero surface charge.

$$\text{Support-OH}_2^+ \; \underset{-H^+}{\overset{+H^+}{\rightleftharpoons}} \; \text{Support-OH} \; \rightleftharpoons \; \text{Support-O}^- + H^+$$

$$\longleftarrow \text{———— PZC ————} \longrightarrow$$

low pH   *high pH*

SUPPORTED METAL OXIDES   Chapter 4

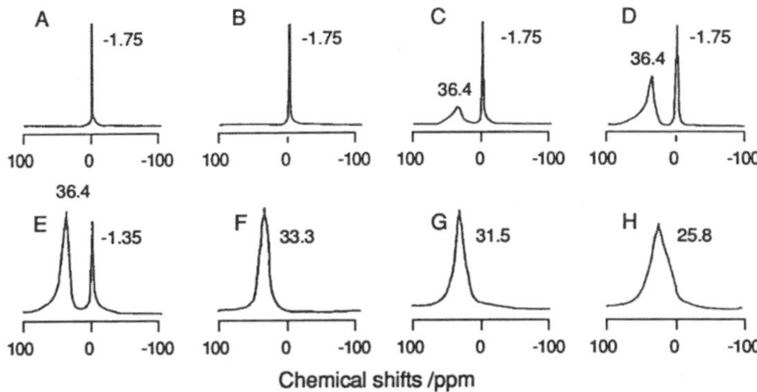

Figure 4.7 ^{95}Mo NMR spectra of 0.1 M molybdenum solution at different pH values: (A) 9.31, (B) 7.61, (C) 6.10, (D) 5.86, (E) 5.73, (F) 5.26, (G) 4.64, and (H) 3.43.

At low pH, the surface hydroxyls of the support are protonated and positively charged, and at high pH, the surface hydroxyls are deprotonated and negatively charged. Consequently, the aqueous film equilibrates at the point of zero charge, where there is a net cancellation of the surface charge. The net pH at PZC is a function of the specific oxide support, the specific metal oxide overlayer, and the surface coverage of the overlayer.[30] Hence, the molecular structures of the metal oxide overlayers under ambient conditions follow the metal oxide aqueous chemistry as a function of the net pH at PZC. From this it is possible to predict the molecular structures of these metal oxide overlayers under ambient conditions. For a support such as MgO, which possesses a pH of 11 at its PZC, only MoO_4 species are present; and for a support such as SiO_2, which possesses a pH of 4 at its PZC, only Mo_7O_{24} species are present. The silica-supported molybdenum oxide system has recently been characterized by Raman spectroscopy,[22, 30] EXAFS/XANES,[22] and Mo-95 solid state NMR.[17] All these studies demonstrated that the Mo_7O_{24} cluster was the Mo species present under ambient conditions. This model has also been successful in predicting the influence of impurities, promoters, calcination temperatures, and the specific metal oxide overlayer.[30] Furthermore, this model also explains why the metal oxide molecular structures cannot be controlled by the preparation methods, since all preparations eventually equilibrate at the same net surface pH at PZC when exposed to ambient conditions.[6]

4.6   Supported Metal Oxide Catalysts Under In Situ Conditions

Under in situ conditions, the moisture present on the oxide support surface is desorbed (usually by heating to 200–400 °C), and the resultant structures of the surface metal oxide phases are no longer controlled by the aqueous solution chemistry.[12, 14, 15, 22, 31] Molecular structures formed under in situ conditions are usually

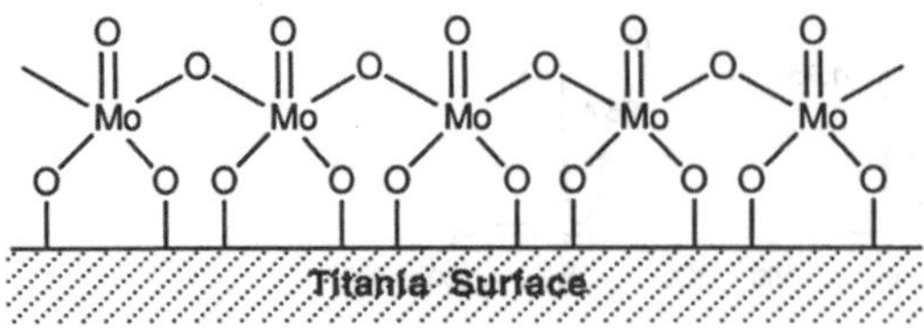

Scheme 4.1 Two-dimensional model for oxidized MoO_3/TiO_2 (6.6 wt %, pH = 4.0).

highly distorted and contain terminal M=O bonds.[22, 31] Surprisingly, essentially the same metal oxide overlayers are formed on different oxide supports[31]; but the specific structures are dependent on the surface metal oxide system being investigated (e.g., Re, Cr, Mo, W, V). For the silica-supported molybdenum oxide system, in situ Raman spectroscopy,[22] EXAFS/XANES,[22] and Mo-95 NMR[17] have revealed the presence of an isolated and highly distorted surface molybdenum oxide species possessing one terminal Mo=O bond (Raman band at approximately 990 cm^{-1}) and four bridging Mo–O–Si bonds. A similar species appears to form on titania supports (see Figure 4.2), but this species is thought to be polymerized and possesses bridging Mo–O–Ti and Mo–O–Mo bonds, as shown below in Scheme 4.1. On support surfaces where polymerized surface species are present, it is highly probable that isolated and polymerized surface metal oxide species coexist. The similar surface metal oxide structures which form on many oxide supports further reveal that these surface metal oxide species have a preferential coordination which cannot be controlled by preparation methods.

4.7 Catalysis and Structure–Reactivity Relationship

The catalytic properties of the two-dimensional metal oxide overlayers are beginning to receive much attention. The influence of surface molybdenum oxide coverage on titania for the methanol oxidation reaction is shown in Figure 4.8. Recall from above that monolayer coverage for this system corresponds to approximately 6.6% MoO_3. The rate of methanol oxidation per Mo site per second, the turnover number (TON), increases slightly with coverage (by approximately a factor of two). This slight increase may be due to either a change in the ratio of isolated to polymerized surface Mo species (becoming more polymerized at higher coverages) or lateral interactions in the molybdenum oxide overlayer. The sharp decrease in the TON above monolayer coverage for the 15.2% sample is caused by the formation of MoO_3 crystallites, which apparently are almost inactive for this reaction. Thus, the surface molybdenum oxide species is significantly more active than the crystalline MoO_3 phase. Similar results have been found for other surface metal oxides phases on oxide supports (i.e., V and Cr).[31]

SUPPORTED METAL OXIDES   Chapter 4

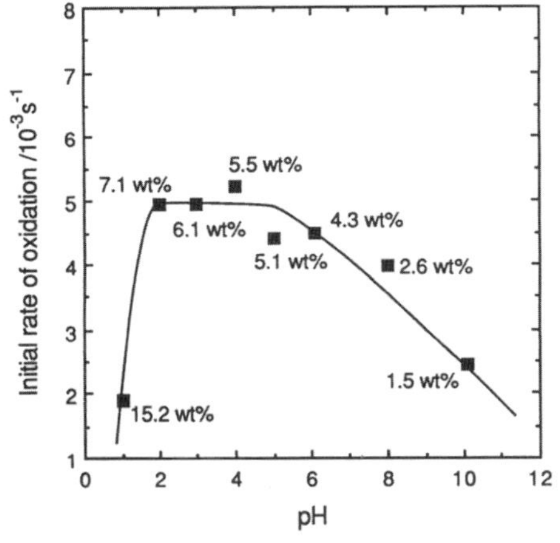

Figure 4.8 Initial rate of methanol oxidation (TON) as a function of Mo loadings (pH of impregnating solution) at 473 K. Numbers in figure represent Mo loading (wt %) at different pH values of impregnating solution.

The influence of different oxide supports on the reactivity of the surface molybdenum oxide species has also been investigated.[31] As already mentioned above, essentially the same surface molybdenum oxide species form on the oxide supports under in situ conditions. However, the TON for methanol oxidation over supported molybdenum oxide was found to be a very strong function of the oxide supports, varying by several orders of magnitude. Reducible oxide supports such as titania (which readily undergoes bulk reduction) and zirconia (which readily undergoes surface reduction) were found to be very active, whereas irreducible oxide supports such as silica and alumina were found to have very low activity. This support effect is thought to be caused by either an increase in the reactivity per surface Mo site or an increase in the number of active sites or a combination of both. Furthermore, the TON was found to correlate with the reducibility of the surface molybdenum oxide species in TPR experiments. Additional experiments are currently being conducted to address this interesting phenomenon. The similar results found for other surface metal oxide phases on different oxide supports (V, Cr, and Re)[28, 31] substantiate that this phenomenon is related to the oxide support and not the surface metal oxide overlayers.

The above discussion is limited only to the methanol oxidation reaction, and the above observations may be strongly tied to that particular catalytic reaction. For example, for the oxidation of *o*-xylene to phthalic anhydride over titania-supported vanadia catalysts, it is critical to have at least monolayer coverage, since the exposed Ti sites decompose the desired products.[32] For such a reaction system, the surface metal oxide coverage has a pronounced impact on the selectivity and activity of the catalyst. However, methanol oxidation selectivity is not a function of coverage, since the exposed Ti sites are not active for this reaction. The TON for the methanol oxidation reaction is a very weak function of surface metal oxide coverage up to monolayer coverage, but the TON for the selective catalytic reduction of NOx by ammonia appears to increase by approximately an order of magnitude up

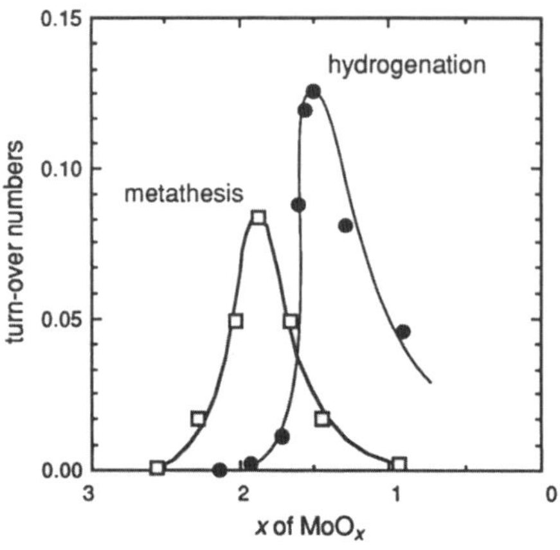

Figure 4.9 Catalytic activities of hydrogenation and metathesis on reduced MoO_3/TiO_2 catalyst (6.6 wt %, pH = 4.0) at different extents of reduction.

to monolayer coverage.[33] This difference is probably due to the facts that methanol oxidation is a unimolecular reaction that occurs on the active surface metal oxide site and that NOx reduction with NH_3 is a bimolecular reaction that requires the participation of the active surface metal oxide site as well as adjacent surface hydroxyls. The specific oxide support also appears to have an important effect on other catalytic reactions, but quantitative TON data are not currently available for comparison.[34]

The above discussion on catalysis of supported metal oxide catalysts focused on partial oxidation reactions, in which the catalyst was exposed to a partial pressure of oxygen. In such systems, the surface metal oxide phases are reduced and reoxidized by the reaction medium (the redox process). Surface metal oxide catalysts are also used to catalyze reactions where a partial pressure of oxygen is not present and the surface metal oxide species possess a reduced state. Examples of such reactions are polymerization, hydrogenation, and metathesis reactions. Refer to the section earlier in this chapter, "Surface Chemistry of Supported Metal Oxides," for additional information about the surface chemistry of such reduced surface metal oxide overlayers. The catalytic activities of titania-supported molybdenum oxide— containing monolayer coverage—for these reactions are shown in Figure 4.9 as a function of the extent of molybdenum oxide reduction. Note that for both of these reactions the TON is a function of the extent of reduction of the surface molybdenum oxide phase, with neither reaction proceeding on the fully oxidized surface. The metathesis reaction exhibits a maximum in TON at $MoO_{1.9}$, and the hydrogenation reaction exhibits a maximum in TON at $MoO_{1.5}$. The reason for this difference is that the metathesis reaction requires two coordinative unsaturated sites (Mo^{4+})—sites where oxygen ligands are removed and chemisorption can be accommodated—so that the two olefins can coordinate to the Mo site and undergo metathesis (see Scheme 4.2). On the other hand, the hydrogenation reaction requires three coordinative unsaturated sites (Mo^{3+}) so that two hydrogen atoms and one olefin can coordinate to the Mo site and undergo hydrogenation (see Scheme 4.2).

     SUPPORTED METAL OXIDES   Chapter 4

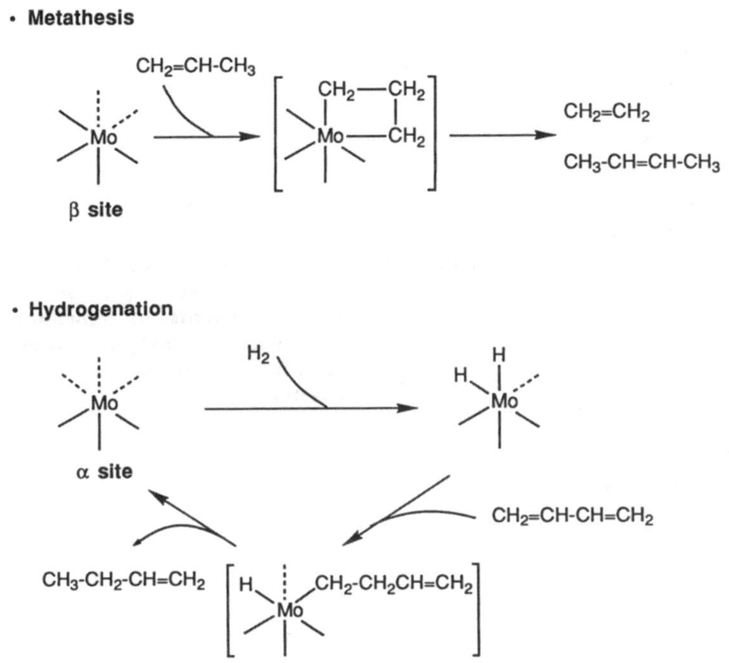
• Metathesis
CH2=CH-CH3
Mo
β site
CH2—CH2
Mo—CH2
CH2=CH2
CH3-CH=CH-CH3
• Hydrogenation
Mo
α site
H2
H
H
Mo
CH2=CH-CH=CH2
CH3-CH2-CH=CH2
H
Mo
CH2-CH2CH=CH2

Scheme 4.2 Reaction pathways of metathesis of propene and hydrogenation of 1,3-butadiene on reduced MoO_3/TiO_2 catalyst. Dashed lines represent coordinatively unsaturated sites (CUS).

Note that the TON of these two reactions over the reduced catalysts correlate with the two NO states observed in the TPD spectra for this system (see Figures 4.6 and 4.10) and are consistent with the requirement of two different reduced sites for these two reactions. Thus, the specific demands of each reaction are different, and different extents of reduction are required for optimum catalytic performance.

4.8 Summary

Supported metal oxide catalysts consist of a two-dimensional metal oxide overlayer on a high surface area oxide support. For some applications the surface metal oxide phase is fully oxidized, but for others the surface metal oxide phase is partially reduced. Among the most critical properties which characterize such complex metal oxide systems are:

- surface coverage of the supported metal oxide phase
- BET surface area
- local structural environment(s) of the surface metal oxide
- oxidation state(s) of the surface metal oxide

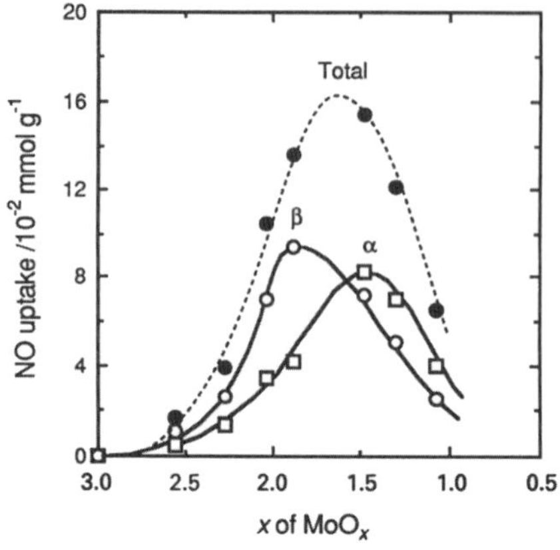

Figure 4.10 NO chemisorption on MoO_3/TiO_2 catalyst (6.6 wt %, pH = 4.0) at different extents of reduction. The curves of α and β were calculated from the peak area of TPD spectra.

- reducibility of the surface metal oxide

- surface chemistry

- catalysis.

Significant progress has been made recently in characterizing the fully oxidized surface metal oxide phases by applying sophisticated molecular spectroscopies (Raman, IR, NMR, and EXAFS/XANES) to this class of catalytic materials. In contrast, very little is still known about the reduced surface metal oxide phases, because the molecular spectroscopies are not so compatible with the distribution of oxidation states that may exist in these catalysts. It is anticipated that much progress will be made in characterizing supported metal oxide catalysts in the coming years.

References

1   C. L. Thomas. *Catalytic Processes and Proven Catalysts*. Academic Press, New York, 1970.

2   *Preparation of Catalysts I–V*. In series *Studies in Surface Science and Catalysis*. (B. Delmon and J. T. Yates, Eds.) Elsevier, Amsterdam, 1976, 1979, 1983, 1987, 1991.

3   L. Wang and W. K. Hall. *J. Catal.* **77**, 232, 1982.

4   G. C. Bond, S. Flamerz, and L. van Wijk. *Catal. Today.* **1**, 2286, 1987.

5   J. Haber, T. Machej, and R. Grabowski. *Solid State Ionics.* **32/33**, 887, 1989.

6   T. Machej, J. Haber, A. M. Turek, and I. E. Wachs. *Appl. Catal.* **70**, 115, 1991.

7   C. Cristiani, P. Forzatti, and G. Busca. *J. Catal.* **116**, 586, 1989.

8   F. Roozeboom, M. C. Mittelmjier-Hazeleger, J. A. Moulijn, J. Medema, V. H. J. de Beer, and P. J. Gellings. *J. Phys. Chem.* **84**, 2783, 1980.

9   K. Segawa and W. K. Hall. *J. Catal.* **76**, 133, 1982.

10   D. S. Kim, Y. Kurusu, I. E. Wachs, F. D. Hardcastle, and K. Segawa. *J. Catal.* **120**, 325, 1989.

11   I. E. Wachs, C. C. Chersich, and J. H. Hardenburgh. *Appl. Catal.* **13**, 339, 1985.

12   I. E. Wachs, F. D. Hardcastle, and S. S. Chan. *Spectroscopy.* **1** (8), 30, 1986.

13   F. W. Lytle, R. B. Greegor, and E. C. Marques. "Catalyst Research with X-ray Absorption Spectroscopy," In *Proc. 9th Intern. Congr. Catal., Vol. 5.* (M. J. Philips and M. Ternan, Eds.) Chemical Institute of Canada, Ottawa, 1988, pp. 54–84.

14   F. D. Hardcastle and I. E. Wachs. *J. Raman Spectroscopy.* **21**, 683, 1990.

15   H. Eckert and I. E. Wachs. *J. Phys. Chem.* **93**, 6796, 1989.

16   E. C. DeCanio, J. C. Edwards, T. R. Scalzo, D. A. Storm, and J. W. Bruno. *J. Catal.* **132**, 498, 1991.

17   G. Haller. Personal communication.

18   H. Eckert. Personal communication.

19   J. C. Edwards, R. D. Adams, and P. E. Ellis. *J. Am. Chem. Soc.* **112**, 8349, 1990.

20   J. A. Horsley, I. E. Wachs, G. H. Via, J. M. Brown, and F. D. Hardcastle. *J. Phys. Chem.* **91**, 4014, 1987.

21   F. D. Hardcastle, I. E. Wachs, J. A. Horsley, and G. H. Via. *J. Mol. Catal.* **46**, 15, 1988.

22   M. deBoer, A. J. van Dillon, D. C. Koningsberger, J. W. Geus, M. A. Vuurman, and I. E. Wachs. *Catal. Letters.* **11**, 227, 1991.

23   S. Srinivasan and A. K. Datye. *Catal. Letters.* Forthcoming in 1992.

24   L. R. Wallenberg, M. Sanati, and A. Andersson. *J. Catal.* **126**, 246, 1990.

25   D. Smith and I. E. Wachs. Unpublished results.

26   S. Srinivasan, A. K. Datye, M. Hapden-Smith, I. E. Wachs, G. Deo, J. M. Jehng, A. M. Turek, and C. H. F. Peden. *J. Catal.* **131**, 260, 1991.

27   J. Kijenski and A. Baiker. *Catal. Today.* **5**, 1, 1989.

28   G. Deo and I. E. Wachs. *J. Catal.* **129**, 307, 1991.

29   J. M. Stencel. *Raman Spectroscopy for Catalysis.* Van Nostrand Reinhold, New York, 1990.

30   G. Deo and I. E. Wachs. *J. Phys. Chem.* **95**, 5889, 1991.

31   I. E. Wachs, G. Deo, D. S. Kim, M. A. Vuurman, and H. Hu. "Molecular Design of Supported Metal Oxide Catalysts." To appear in *Proc. 10th Inter. Congr. Catalysis.* Budapest, 1992.

32 I. E. Wachs, S. S. Chan, R. Y. Saleh, and C. C. Chersich. *Appl. Catal.* **15**, 339, 1985.

33 G. T. Went, L. Leu, R. R. Rosin, and A. T. Bell. *J. Catal.* **134**, 492, 1992.

34 K. Hauffe and H. Raveling. *Ber. Bunsenges. Phys. Chem.* **84**, 912, 1980.

                                                                          SUPPORTED METAL OXIDES   Chapter 4

5

Bulk Metal Sulfides

MICHEL DAAGE

Contents

5.1 Introduction

Sulfide catalysts have been used for many years in the petroleum and chemical industries. Early in this century, sulfur-containing compounds were associated with the poisoning of metallic catalysts used in the conversion of coal, petroleum, and resids. Initially developed for the hydrogenation of coal, sulfide catalysts came to be used in petroleum processing just prior to and during World War II. Since then, sulfides have been used for a wide range of catalytic reactions.

Compared to other catalysts, transition metal sulfides (TMS) fall in a special category due to their exceptional resistance to poisons. In fact, sulfur compounds—the most common poison of metallic and oxide catalysts—do not decrease the catalytic activity of TMS, but instead are needed to maintain high catalytic activity. Sulfide catalysts are also very resistant to carbon deposition, which is why they are well-suited for use in converting resids. Arsenic, nickel, and vanadium, which are contained in heavy petroleum fractions, are some of the very few substances which cause significant deactivation of TMS.

TMS are prominently used for hydroprocessing petroleum feedstocks. Because of obvious economic considerations, bulk TMS is used less commonly than the supported TMS analog, and is somewhat limited to the production of higher value-

"

added chemicals. The development of supported TMS catalysts has been the focus of the petroleum industry. With the exception of processes requiring a bifunctional catalyst (hydrocracking for example), the ability to catalyze a given reaction is specific to the sulfide phase and does not depend on the presence of the support. In fact, the interaction between the pure sulfide phase and the support, though beneficial for dispersion, may sometimes result in the partial loss of critical bulk properties.

The major reactions catalyzed by TMS are hydrogenation of olefins, ketones, and aromatics; hydrodesulfurization (HDS); hydrodenitrogenation (HDN); hydrodemetallization (HDM); dealkylation; and ring opening of aromatics. Other uses of TMS catalysts include reforming, isomerization of paraffins, dehydrogenation of alcohols, Fischer–Tropsch and alcohol synthesis, hydration of olefins, amination, mercaptan and thiophene synthesis, and direct coal liquefaction. More esoteric applications can also be found in polymer synthesis and inorganic synthesis (the synthesis of silanes and metal hydrides). A more comprehensive list of TMS catalyzed reactions has been compiled by Weisser and Landa.[1]

5.2 Preparation of Bulk TMS Catalysts

TMS catalysts are generally prepared by direct sulfidation of a metal salt (oxide, halide, etc.) or by decomposition of a sulfur-containing precursor. Mixed metal sulfides—promoted catalysts—can also be obtained by direct synthesis or by subsequent impregnation of the promoter metal onto the pure sulfide. The resulting precursor is generally dried and then converted to a sulfide catalyst by thermal treatment, according to specific conditions. This last step is often the most critical and is specific in many cases. The following paragraphs describe a few general synthetic procedures that are known for providing reliable precursors.

Binary Sulfides

Most binary sulfides can be obtained by direct sulfidation of oxides and chlorides in the presence of either hydrogen sulfide or a mixture of 10–15% hydrogen sulfide in hydrogen, but low temperature precipitation is often preferred. A general synthetic method has been reported by Chianelli et al.[2] for the preparation of Group IVB, VB, and VIB binary sulfides. The low temperature metathetical reactions are convenient and allow for good control over parameters such as particle size and composition. The general equation is

$$MX_n + (0.5n)A_2S \longrightarrow MS_{0.5n} + nAX$$

where M is the transition metal ion, X is the the salt anion (Cl^-, carboxylate, etc.), and A is an alkali-like cation (Li^+, Na^+, NH_4^+, etc.). The materials obtained using this reaction are generally amorphous and have high surface area. Solid which are poorly crystalline to crystalline are then obtained by thermal treatment under flowing inert gas, hydrogen, or H_2S/H_2.

Another common synthesis of Group VIB sulfides consists in the preparation of thiosalts, such as $(NH_4)_2MoS_4$ and $(NH_4)_2\,WS_4$. These salts are easily obtained by bubbling hydrogen sulfide in a solution of ammonium molybdate or a solution of tungstate in ammonium hydroxide at room temperature. The thiosalts are converted to the corresponding sulfides by thermal decomposition, as described above. Similarly, tetraalkyl ammonium thiosalts have been synthesized for Mo, W, and Re.[3]

Group VIII sulfides are obtained by the low temperature precipitation method described above. However, a preferred synthesis involves the direct sulfidation of ammonium hexachlorometallates with hydrogen sulfide. A good example is found in the synthesis of RuS_2 and Rh_2S_3.

Finally, highly crystalline sulfides (or single crystals) are generally obtained from the elements by using a vapor transport technique in the presence of a trace amount of halides. Such materials are often used as models or references for the characterization of the catalysts.

Mixed Metal Sulfides

Mixed metal sulfides are generally prepared by one of the four following methods:

- comaceration

- homogeneous sulfide precipitation

- mixed metal salts or organometallic clusters decomposition

- impregnation of a binary sulfide.

Most of these techniques were developed for the synthesis of Co/Mo or Ni/Mo catalysts. However, they are also applicable to other systems.

The comaceration method consists in reacting freshly prepared oxides with a solution of ammonium sulfide at a moderate temperature (about 70 °C). The slurry is continuously stirred until it evaporates to dryness. The catalyst is obtained by thermal treatment under vacuum.[4]

Homogeneous sulfide precipitation (HSP) is a type of low temperature coprecipitation in which two or more salts are dissolved prior to the addition of the sulfiding agent. A good use for the HSP method is the synthesis of unsupported Co–Mo sulfide. Here, a solution of cobalt nitrate and ammonium heptamolybdate is poured into a hot solution of ammonium sulfide. The slurry is then evaporated to dryness.[5]

Recently, thermal decomposition of molecular metal–sulfur complexes either start from salts, where both the anion and the cation contain one of the metals, or from molecular complexes, in which the two metals are part of the same molecular sulfur-ligated cluster. $Co(en)_3MoS_4$ and Co–Mo cube structures are classical examples of such precursors[6, 7] These precursors can then be decomposed under various conditions—in an inert atmosphere, hydrogen, H_2S/H_2, or in situ. In contrast with the other methods described above, there is an inherent limitation to this approach in that there is little flexibility in obtaining a given promoter to metal ratio.

The second metal, the promoter, can also be added by a subsequent impregnation of binary sulfide. When a nonreactive promoter precursor—for example, a metal nitrate—is used, it is necessary to resulfide the impregnated sulfide in order to decompose the precursor. Another variation of this method uses reactive promoter precursors that will react with the surface of the binary sulfide. In this case, further treatment of the catalyst may not be required. Effective precursors include metal carbonyls and metal alkyls[8, 9]

Highly crystalline materials or doped single crystals or both can be obtained from the elements by vapor transport techniques[10] or by impregnation of the highly crystalline binary sulfides.[11]

5.3   Bulk Characterization

Structures of TMS and Stable Catalytic Phases

The catalytic activity of TMS catalysts is related to defects in their crystal lattice, particularly at the surface. They are somewhat similar to bulk oxides in that perfect crystals do not achieve high activity. Therefore, it is particularly important to know the phases that are stable in a catalytic environment and the crystal structure anomalies that are frequently encountered. (Halbert et al.[11] detail some of the wide variety of phases in which sulfides exist.) Because sulfide catalysts are high temperature catalysts used mostly in reducing conditions, significant restructuring of TMS phases occurs under catalytic conditions. In situ characterization of TMS materials is generally difficult and limited because of the in situ conditions: high temperature, high pressure, and liquid phase reactions. Therefore, the fresh catalyst and the spent catalyst are generally characterized. The most common catalytic phases are likely to remain the same after exposure to reaction conditions. Exceptions may occur when the high temperature phase is not stable at room temperature. After hydrodesulfurization, the stable phases obtained for binary sulfides are TiS_2,

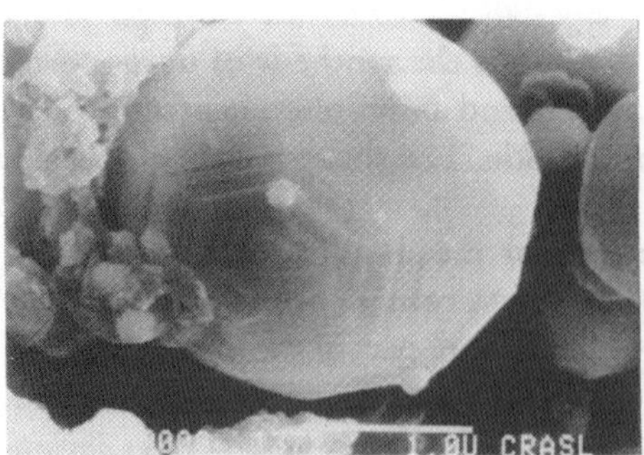
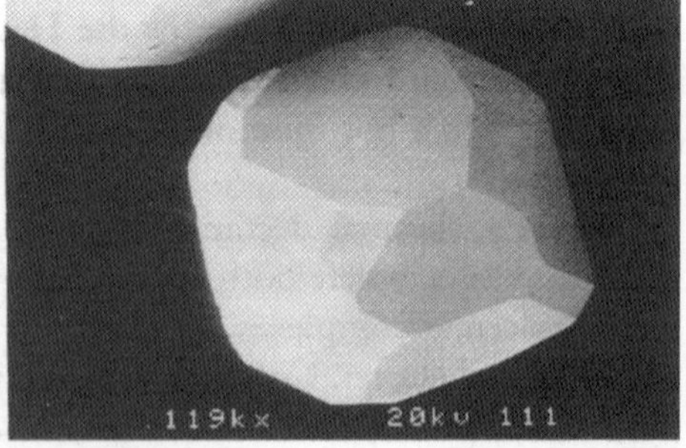

Figure 5.1   **Electron micrographs of isotropic sulfide crystals: (left) Co_9S_8 and (right) RuS_2.**

   BULK METAL SULFIDES   Chapter 5

VS_x, Cr_2S_3, MnS, FeS_{1-x}, Co_9S_8, Ni_2S_3, NbS_2, MoS_2, RuS_{2-x}, Rh_2S_3, PdS_x, TaS_2, WS_2, ReS_2, OsS_x, IrS_x, PtS_{1-x}.

The identification and composition of these phases are obtained using conventional techniques such as X-ray diffraction (XRD) and elemental analysis. However, good catalysts are often poorly crystalline and produce broad X-ray diffraction peaks, making the interpretation of diffraction patterns a complex task that requires particular care.

For the binary phases, there are two classes of TMS structures, isotropic sulfides and layered sulfides. Isotropic sulfides (see Figure 5.1) include Group VIII sulfides and manganese sulfides. Pyrites, pentlandites, and heazlewoodites are the common structures. Layered sulfides (see Figure 5.2) include Groups IV through VII sulfides. Their main structural feature is the strong anisotropy of the crystallites. A characteristic structure of this group is molybdenite.

When a second or third metal is present, the identification of phases becomes more complicated. When both components belong to the isotropic class, solid solutions that crystallize in a cubic pyrite structure may be obtained, as is observed for $Co_{1-x}Ru_xS_2$, $Co_{1-x}Rh_xS_2$, and $Rh_{1-x}Ru_xS_2$[13].

For materials that contain at least one element that crystallizes in a layered phase, few well-defined mixed compounds exist that correspond to intercalates such as $CoMo_2S_4$. Because these materials cannot be prepared in a dispersed form, their catalytic activity is low. Recently, Chevrel phases have been used for HDS,[14] but the necessary high temperature synthesis is a major obstacle in preparing active catalysts. In general, poorly crystalline mixed-metal materials do not correspond to an intercalated phase, and under catalytic conditions, phase segregation is severe. The XRD pattern is characteristic of a mixture of two corresponding binary phases. It is very difficult to find any evidence of a mixed phase in the diffractogram. This is generally a result of the formation of the mixed phase at the surface of the bulk sulfide, which then acts as the support. Auger analysis and analytical electron microscopy (AEM) have been used on cobalt-doped MoS_2 crystals after sputtering.[10, 15] Clear evidence that the Co promoter is concentrated near the edge plane of the crystal is obtained through a comparison of the Co/Mo ratio obtained along the c-axis and a-axis of the crystal. In order to determine the precise structure of

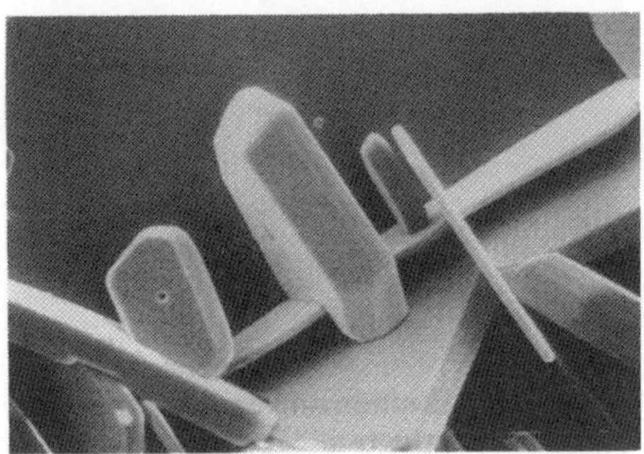

Figure 5.2   **Electron micrograph of an anisotropic sulfide—MoS$_2$ platelets.**

the promoter, more sophisticated techniques are required, which are discussed later in this chapter in the section, "Surface Composition."

Morphology, Particle Size, and Surface Area

The morphology, particle size, and surface area of TMS are generally obtained using conventional methods such as electron microscopy, X-ray diffraction line-broadening, and nitrogen BET. In contrast to the isotropic class, major difficulties are encountered when characterizing materials of the anisotropic class. For example, molybdenum disulfide in a poorly crystalline state develops a morphology of its own, best described as "rag structure." A quick look at an electron micrograph (Figure 5.3) makes clearly evident the difficulty of estimating the particle size. XRD may be used, but the line-broadening will be characteristic of the crystalline order, which depends upon the bending and folding of the layer. When the Debye-Scherrer equation is applied to the 002 diffraction peak, a reasonable value for the c-axis of the crystal is obtained for the stacking of the layers (within a factor of two), but a line-broadening analysis of the 103 or 110 peaks results in values along the a-axis between 70 and 100 Å, though the actual dimension can be as large as several microns.

Extended X-ray fine-structure spectroscopy (EXAFS) measurements have also been used for determining particle size, particularly in supported materials. The average coordination numbers of the metal atoms are generally used for estimating particle size. For isotropic materials, such estimations usually agree well with the results obtained by the preceding methods. In contrast, significant discrepancies can be observed for highly anisotropic materials. For molybdenum disulfide, the attenuation of the Mo–Mo peak in a sample treated at 900 °C is 50% of that observed for a single crystal. Therefore, the corresponding coordination number suggests particle sizes that are much smaller than the 20 000 Å particles estimated by the SEM micrograph (see Figure 5.3).

BET measurements are adequate in all cases and are suitable for surface areas up to 400 m^2/g. After catalytic reaction, most poorly crystalline sulfides have surface areas in the 10–100 m^2/g range. Most of the surface area of the anisotropic class is associated with the basal plane of the layer, which consists of a closely packed surface of sulfur atoms. If an infinite layer of MoS_2 has a surface area of $327 m^2/g$

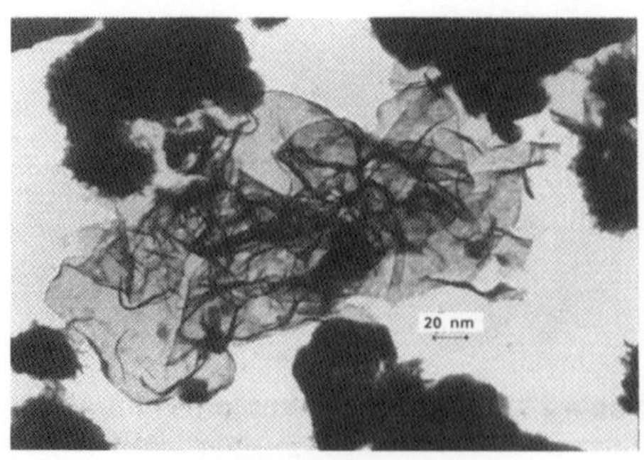

Figure 5.3   **Electron micrograph of the "rag structure" of MoS_2.**

and if n is the number of stacked layers of large particles on the surface, then the surface area of the particles is $327/n$ because the surface area of relatively large particles is inversely proportional to their stacking. Also, the basal plane surface consists of a closely packed surface of sulfur atoms, each bonded to three metal atoms. The stability of the sulfur environment is in fact related to the weak van der Waals interaction between the layers. The basal plane exhibits a low reactivity and is of central interest to lubrication science. In contrast, the edge plane is highly reactive and is of primary importance for developing structure/function relationships in anisotropic systems. There is no easy way of measuring the edge plane area except when single microcrystals are used; the edge surface area is then directly estimated from SEM micrographs.[16] Several attempts have been made to measure edge surface area using chemisorption techniques. This is discussed later in this chapter in the section, "Surface Composition."

Precautions should be taken on measurements for mixed metal systems where anisotropic structures are likely to be observed. The second metal, generally of the first transition metal row, affects the morphology by increasing the stacking while decreasing the diameter of the layers. In cases where severe phase segregation occurs, the second metal sulfide may act as a support, thereby dispersing the catalytic phase. Such morphology changes are generally easily identified by transmission electron microscopy.

Metal Oxidation State and Structural Environment

The oxidation state of the metals in the bulk corresponds to that expected for a given phase. Generally the lattice sulfur is tightly bonded to several metal atoms (most often three) and is very stable; consequently, narrow valency distributions are observed. Other metal oxidation states are sometimes detected, though their concentration is low and they are generally associated with the presence of defects or sulfur vacancies. For example, an electron spin resonance (ESR) signal, attributed to defects such as Mo^{III} or Mo^{V}, can easily be observed in poorly crystalline MoS_2. However, magnetic susceptibility measurements provide evidence that spin coupling due to cation pairing occurs, suggesting that only a fraction of the actual defects are detected by ESR.

In some cases, there is a correlation between the defect concentration and the catalytic activity of the sulfide. However, in contrast with some oxide catalysts in which the lattice oxygen has been identified as a critical component of the catalytic behavior, the bulk atoms of the sulfides do not participate directly in the reaction. The bulk should be considered as a support for the catalytically active surface structure. Therefore, bulk property characteristics do not allow the determination of site densities and turnover frequencies, but can only be used as a relative measurement to compare a family of samples.

Platinum, iridium and osmium should be regarded as special cases, because they form multiphasic materials under reducing conditions. Zero valent metal often coexists with the sulfide phase and is detected by X-ray diffraction. The wider oxidation state distributions can be obtained by X-ray photoelectron spectroscopy (XPS), ESR, MAS-NMR and EXAFS where appropriate. Similarly, palladium undergoes drastic changes during catalysis, with different materials ranging from poorly crystalline PdS to crystalline Pd_4S obtainable.

EXAFS and, in particular, X-ray absorption near-edge spectroscopy (XANES) are used to determine the coordination number of the metals and their local environment. The *average* local structure is deduced by comparing the spectral features of the catalyst with the spectra of well-defined model compounds. Octahedral, trigonal prismatic, tetrahedral, and square pyramidal environments have been reported for the metal atoms.[17] There are limitations in using this technique. In amorphous or poorly crystalline materials, the environment of the metal is generally distorted, which causes the coordination numbers to be lower than expected. This effect is more pronounced for highly anisotropic sulfides because of the folding of their layers. As in the case of MoS_2, the coordination number is then unreliable for estimating particle sizes.

5.4 Surface Composition

Catalytic events such as adsorption and the breaking and forming of bonds are obviously associated with the surface of the solid. Information regarding the composition of the surface is therefore essential in providing a good understanding of a catalyst's behavior. Multiple approaches are generally used, including chemisorption and spectroscopic methods, in particular.

Chemisorption and Molecular Probes

Chemisorption has been applied to numerous catalytic systems including TMS catalysts, providing valuable information on the active surface area or catalytic site densities. However, in order to interpret the results the stoichiometry of the chemisorption must be known. For metal surfaces, dispersion is calculated using one hydrogen atom per metal atom. However, dispersions higher than 100% are not unusual for highly dispersed catalysts with particle sizes below 15 Å. Thus, it may be assumed that there are metal atoms associated with more than one hydrogen. That assumption may also be applied to TMS catalysts. A good example of this is the chemisorption of oxygen on the edge surface of MoS_2. Site densities are calculated using one oxygen atom per molybdenum atom on the edge, though it has been determined that 15 oxygen atoms per edge molybdenum atom are actually consumed. This clearly shows that the oxygen chemisorption is corrosive. However, a linear relationship between the site density and the oxygen chemisorption does exist due to the topotactic nature of the molybdenum disulfide oxidation.[10] Similar relationships can be observed with promoted molybdenum sulfides. Each family of catalysts has its own linear correlation, which cannot be directly compared to that of another family. Recently it has been claimed that low temperature oxygen chemisorption is more reliable, but it also lacks a well-defined stoichiometry. Oxygen chemisorption has been applied to tungsten and rhenium sulfides and also to promoted molybdenum and tungsten sulfides. In the isotropic class, it has been applied only to ruthenium sulfide.

Other molecules that can be used for the characterization of the active surface of TMS catalysts are CO and NO.[18, 19] CO has been seldomly used, whereas NO has been extensively applied to the CoMoS system. Both molecules have the additional advantage of being easily detected by infrared spectroscopy, which provides information relevant in determining which metal atom they bind. However, in ultra-high vacuum conditions, NO can be decomposed at 130 K, which suggests that the interpretation of the data might be more complex due to possible partial oxidation of the surface.[20]

Bases such as ammonia and pyridine can be used to characterize the acidity of the surface. But, the formation of intercalates with layered materials is quite common and may lead to inconclusive results.

Though chemisorption techniques may lead to confusing interpretations regarding the site density, they remain the most convenient way of screening a family of catalysts, thereby limiting the cost and time needed for reactor tests.

Surface Characterization Using Spectroscopic Techniques

The elemental composition and oxidation states of surfaces are most frequently determined using XPS, ultraviolet photoelectron spectroscopy (UPS), Auger electron spectroscopy, and high-resolution electron energy loss spectroscopy (HREELS). However, there are certain limitations in the application of these techniques. In situ characterization cannot be performed, since ultra-high vacuum is needed. Because of the high sensitivity of sulfides to air exposure, the use of external cells for high temperature and sulfiding is mandatory for obtaining good data. With the exception of molybdenite and pyrite, sulfide single crystals are not commercially available and must be synthetically grown by vapor transport. The resulting crystals are small (1–10 mm in diameter), rendering analysis more difficult, particularly when considering sulfides from the isotropic class. Almost none of the surfaces of the isotropic sulfide catalysts have been characterized—with the exception of RuS_2—because of the nearly exclusive use of molybdenum- or tungsten-based catalysts in the petroleum industry. Most surface studies have been performed on molybdenum sulfide and its promoted analogs. Layered sulfides grow crystals easier than isotropic sulfides, though they are very thin. This has led to a great deal of data being collected on the structure of the basal plane by means of the atomic resolution imaging provided by scanning tunneling microscopy (see Figure 5.4). Spatially-resolved electron energy loss spectroscopy (SREELS) has also been used on MoS_2 platelets, showing an enhancement of surface plasmon modes at corners and edges.[21]

However, the characterization of the catalytically active edge plane remains the major challenge to the surface scientist. For example, the edge surface structure in

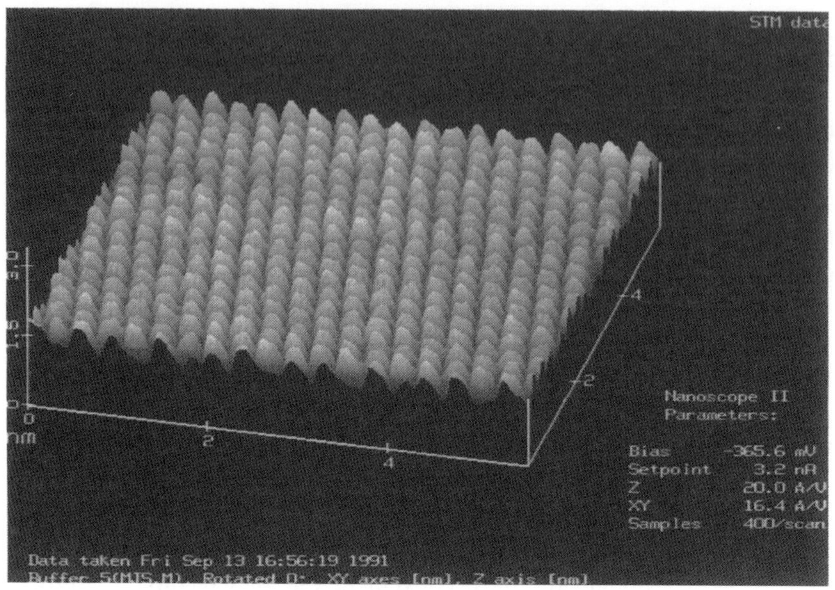

Figure 5.4   Scanning tunneling micrograph of the basal plane of a synthetic MoS_2 single crystal. (This micrograph kindly furnished by Dr. H. Othani.)

molybdenum disulfide depends on the stacking sequence of the layers, with different surfaces existing for the rhombohedral (3R) and hexagonal (2H) structures of the crystal. Each layer will be terminated with the two types of Mo atoms which correspond to the $(\bar{1}010)$ and $(10\bar{1}0)$ planes (see Figure 5.5). The bonding of an organic substrate on the edges is determined principally by the environment of a particular Mo atom on the edge of one single layer; the bonding is therefore not dependent upon the long-range order of a single crystal face. This enables surface techniques to be used on synthetic edge surfaces. A good example is found in the XPS and UPS study of edges created by the lithography of a single crystal of MoS_2 (see Figure 5.6). It was found that the Mo atoms on the edges have a lower oxidation state (less than 4) than that of the bulk (see Figure 5.7). Another surface science approach consists in cleaning an edge surface of a natural crystal, which is generally thicker but does contain some metallic contaminants. An HREELS study of CO adsorbed on a reduced edge surface has been carried out using this approach. Temperature-programmed desorption (TPD) and temperature-programmed reduction (TPR) provide some additional information on the reducibility of the different types of sulfur atoms and their ability to form anionic vacancies. An experiment using these two methods[22] showed that, for MoS_2, terminal sulfurs are removed below 200 °C, whereas bridged sulfurs react between 200 and 500 °C. Above 500 °C the triply bound sulfur becomes unstable and the crystal structure collapses. Scanning Auger spectroscopy and analytical electron spectroscopy have also been used to characterize the edge surface of promoted crystals.

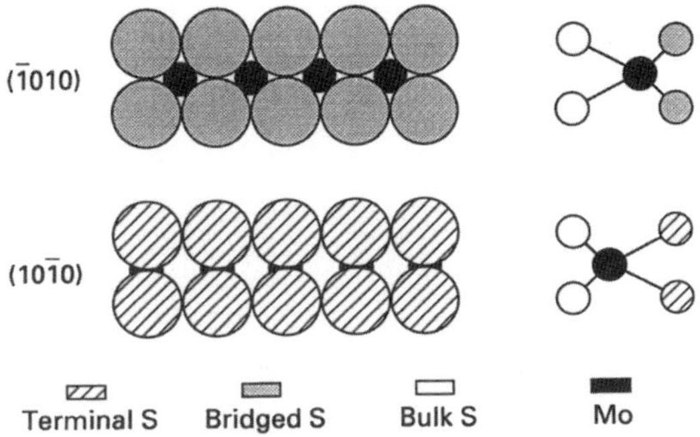

Figure 5.5 Schematic representation of the two edge planes of an MoS₂ layer.

Also, the defects located on the edge planes have well-defined optical characteristics; a midgap optical absorption similar to that observed for other semiconductors provides a direct correlation between the electronic structure of MoS_2 and its catalytic properties. The infrared optical absorption is determined using photo-thermal deflection spectroscopy (PDS).[16] This is a calorimetric technique that directly measures the deflection of a laser beam. It is insensitive to optical scattering, which makes it ideal for powders. Catalytically active defects would be expected to have energy levels between the conduction and valence bands of MoS_2 and should absorb photons with below-bandgap energies. In fact, a set of Xα calculations[23] modeling different types of sulfur vacancies that can occur at the edge surface did show that the allowed optical transitions fall into the observed energy ranges below 1.2 eV. This result suggests that the sulfur vacancies are responsible for the optical absorptions measured for the edge planes. PDS is particularly useful since it is the only spectroscopy that shows the correlation of the catalytic properties for a wide range of MoS_2 dispersions (single crystal to poorly crystalline powders). However, it cannot be used for conductors, such as cobalt-promoted MoS_2.

NMR, Mössbauer spectroscopy, and EXAFS can also be used as surface-sensitive techniques.[24–27] This makes them particularly useful for characterizing the structure of the promoter atoms of a layered TMS catalyst, since such atoms are

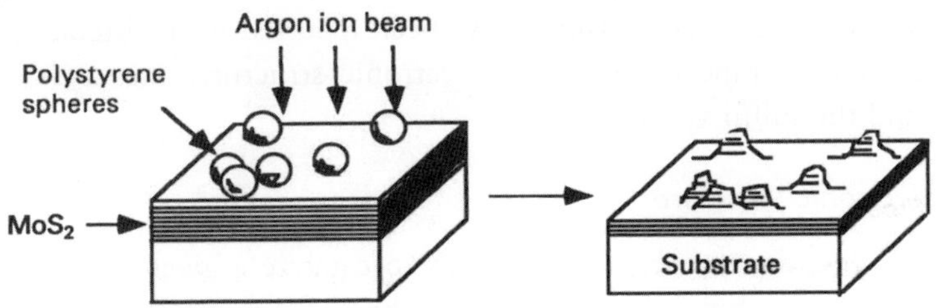

Figure 5.6 Lithographic preparation of MoS₂ edge surfaces.

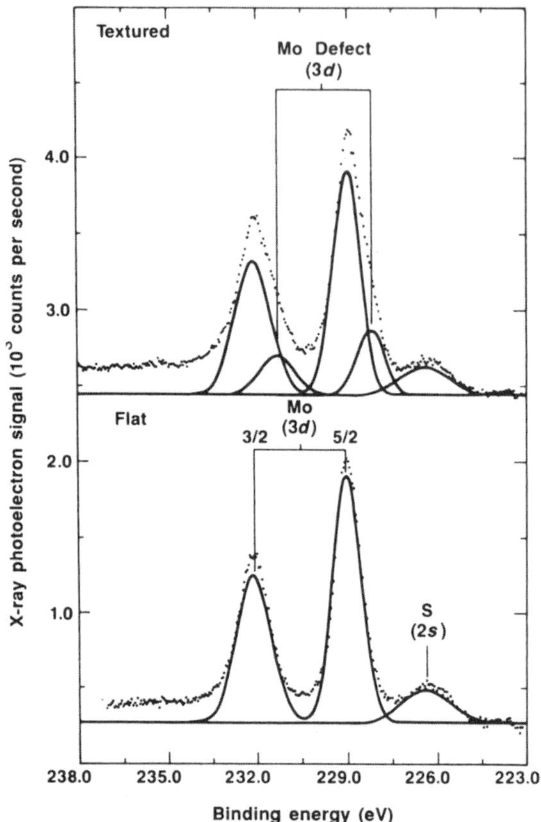

Figure 5.7   X-ray photoemission spectra of MoS$_2$ before and after texturing.

located nearly exclusively on the edge surface. Surface sulfur structure can also be analyzed by using selenium as a tracer atom, allowing characterization by solid state NMR.[28]

5.5   Structure–Function Relationships

Considerable progress has been made in characterizing TMS materials, and much effort has been expended in an attempt to understand the basis for their catalytic activity and selectivity. Numerous relationships have been reported in the literature that highlight the particular importance of the electronic structure, the crystallographic structure, and the sulfur vacancies.

Importance of the Electronic Structure

Periodic effects, which describe the ability of a TMS to catalyze a given reaction, form the underpinning for any fundamental understanding of the electronic structure. These effects were first measured for the desulfurization of dibenzothiophene,

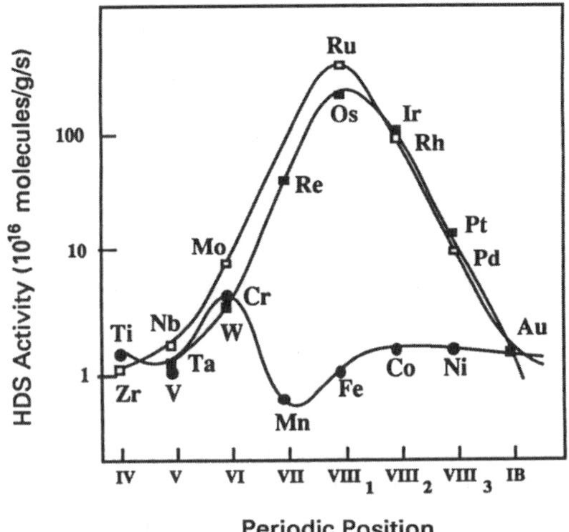

Figure 5.8   Periodic trend of the binary sulfides for the hydrodesulfurization of dibenzothiophene.

where a typical "volcano" plot was seen to emerge between the activity and the periodic position[29] (see Figure 5.8). Group VIII TMS's such as Ru, Rh, Os, and Ir were the most active. Similar trends were later observed for hydrogenation reactions, HDN, and the hydrotreating of a heavy gas oil.[30] These trends are of fundamental importance because they emphasize the importance of the $4d$ and $5d$ electrons in catalyzing these reactions. In this respect, the TMS catalysts resemble the metal catalysts. A theoretical insight for understanding these correlations is found in the calculated bulk electronic structures of the first and second row TMS's. There are three electronic factors that appear to be related to catalytic activity: the electrons in the highest occupied molecular orbital (HOMO), the degree of covalency of the metal–sulfur bond, and the metal–sulfur bond strength. These factors incorporate into an activity parameter, A_2, which correlates well with the periodic trends.[31] The effect of the metal–sulfur bond is supported experimentally by the trend.[31] between catalytic activity and the heat of formation of the binary sulfide. High-activity catalysts have a heat of formation of 30–50 kcal/mol. The metal–sulfur bond should be strong enough for the sulfur-containing molecule to bind to the metal, but weak enough for H_2S to desorb.[32]

The promotional effect of a first row transition metal on the activity of molybdenum-based catalysts correlates with a similar parameter, suggesting an interaction between the Mo $4d$ electrons and the $3d$ electrons of the first row transition metal.[33] This interaction is theoretically described by electronic transfer between the two metals. For example, Co and Ni act as promoters by increasing the number of electrons in the HOMO from 2 in MoS_2 to 3 and 4 in CoMo and NiMo systems, respectively. By contrast, it is found that Cu acts as a poison by abstracting one electron from the HOMO. No promotion or inhibition effects were found for the remaining first row metals. That the effect of the promoter is electronic (chemical) is also supported by the relation of the synergic pairs with the binary sulfides through the average heat of formation. The Co or Ni promoted molybdenum catalysts have an average heat of formation near 40 kcal/mol.[32] An upper limit for the electronic promotion factor has been estimated for MoS_2 microcrystals, which do not undergo any morphological changes when treated with a low valent promoter such as cobalt carbonyl.[11] For cobalt, the electronic factor is 3.3. This factor is common to all morphologies of the Mo/S system, since only a single pair of Co and Mo atoms are involved in the electronic promotion.

Effect of the Crystallographic Structure

As mentioned in the preceding sections, the TMS catalysts belong to two major classes defined by the isotropic/anisotropic character of their crystallographic structures. Clear evidence of this is found in the correlation of catalytic activity with the surface area, or the lack thereof, as illustrated in Figure 5.9 for RuS_2 and MoS_2. The isotropic nature of RuS_2 is clearly demonstrated in the linear correlation between activity and surface area. RuS_2 is an example of a well-behaved catalyst system, unlike MoS_2, in which the anisotropy plays a large role. In contrast, linear relationships between the activity for dibenzothiophene (DBT) HDS and oxygen chemisorption are observed for both materials (see Figure 5.10). This is a clear indication that the chemisorption is at least proportional to the density of catalytic sites. Still, the relationship cannot be used to measure concentration without a precise knowledge of the stoichiometry of the chemisorption process. For RuS_2, the stoichiometry can be estimated from the slope of the linear correlation found between the oxygen chemisorption and the surface area, resulting in a value of about one oxygen

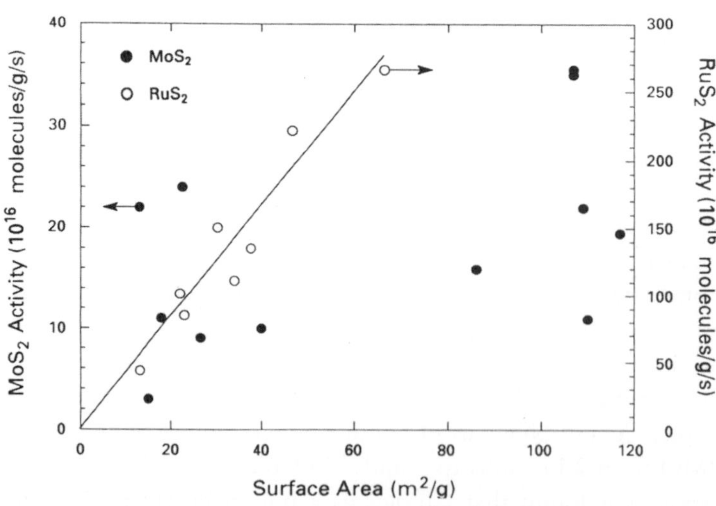

Figure 5.9   **Relationship between activity and surface area for MoS$_2$ and RuS$_2$.**

atom for every two surface Ru atoms. This suggests that a corrosive process is not likely to be associated with the chemisorption. For MoS_2, such a stoichiometric estimation is more difficult. Since the oxygen accumulates on the edges of the crystallites, a knowledge of the edge surface area is critical. As of now, the only simple way of obtaining such information is to geometrically estimate the edge surface area of well-defined microcrystals from SEM micrographs. The catalytic activity of microcrystals having an average dimension of 1.7 μm has been measured,[16] and the turnover frequency for the HDS of DBT has been determined to be 0.079 molecule/site/s. Since the optical defects measured by PDS are of the same nature regardless of the crystallinity of the sample, this turnover frequency is in fact applicable to any MoS_2 catalysts. Consequently, a direct correlation between the edge Mo density and the chemisorption of oxygen is established (see Figure 5.11). The slope of the linear relationship clearly shows that the chemisorption at room temperature is corrosive and that 15 oxygen atoms per edge Mo atom are consumed.

Numerous other correlations of characteristics with activity have been reported for MoS_2, including ESR spin density, magnetic susceptibility, and other chemisorptions.

The promotion of bulk binary sulfide is almost exclusively limited to the promotion of molybdenum and tungsten by a first row transition metal. The effect of this structural promotion is always associated with electronic promotion. One approach for quantifying this consists in estimating the promotion factor obtained for different morphologies and structures of Mo/S systems that convert to the catalytically active Co/Mo/S system. For example, when different MoS_2 microcrystals and

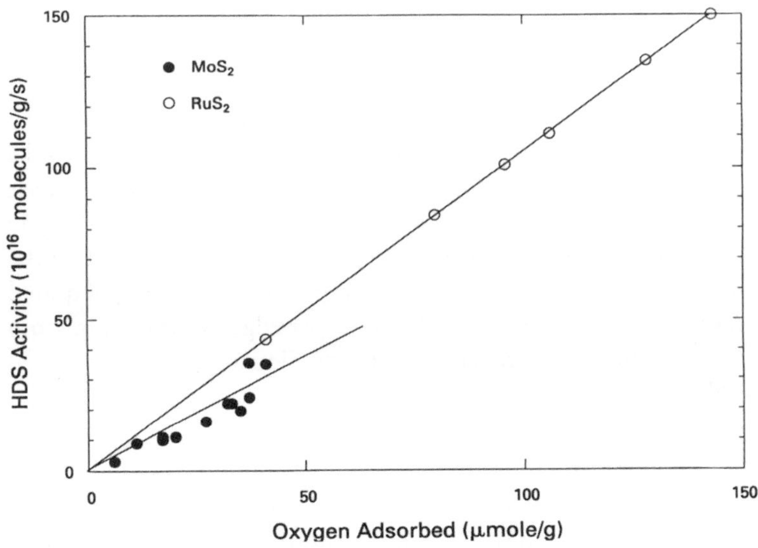

Figure 5.10 **Oxygen chemisorption correlation with activity for MoS₂ and RuS₂.**

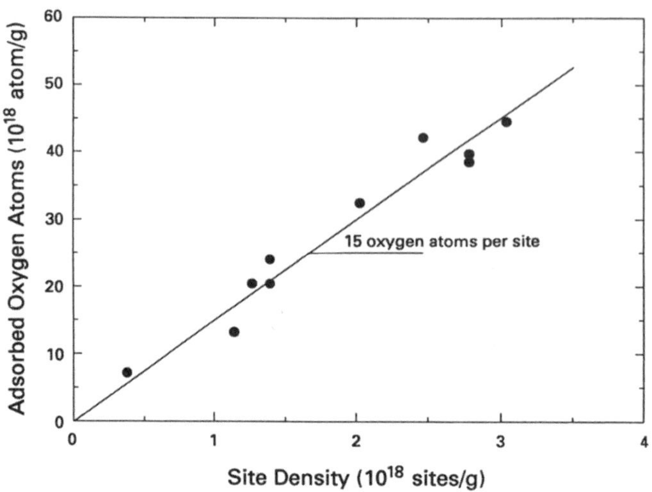

Figure 5.11   The stoichiometry of oxygen chemisorption for MoS$_2$.

poorly crystalline powders are promoted by reaction with cobalt carbonyl,[11] the promotion factor for the HDS of DBT varies between 3.3 and 16. Once corrected for the contribution of the electronic factor, however, the promotion factor lies between 0 and 4.3. The highest values are always obtained for materials where significant morphological changes (observable using TEM) occur upon promotion. It is believed that weak bonds and associations, which prevail in highly disordered materials, are broken, resulting in the formation of smaller particles and hence more catalytic sites. That is the effect responsible for the wide range of promoter to molybdenum ratios that are presented in the literature, where values ranging from 0.1 to 0.8 have been claimed and numerous relationships have been established for different reactions. It is clear, however, that the close association of Co with molybdenum is critical. A good correlation exists between activity and the amount of Co present in the MoS$_2$-like "CoMoS" phase, as measured by Mössbauer spectroscopy.[25] (The "CoMoS" phase generally coexists with the binary phases, its amount and dispersion being highly dependent upon the mode of preparation.) In this phase, the cobalt is located on the edge of a small slab of MoS$_2$, allowing the formation of the Co–Mo pair required for electronic promotion. The number of cobalt atoms per edge molybdenum atom needed for maximum catalytic activity remains unknown. Similar phases are believed to exist for Fe/Mo, Ni/Mo, Co/W, and Ni/W.

Effect of the Sulfur Vacancies

The formation of sulfur vacancies is somewhat related to the electronic effect, in that a relatively narrow range of metal sulfur bond strength is required for high catalytic activity. This gives rise to the general belief that sulfur vacancies are critical.

However, some recent organometallic studies suggest that the catalytic reaction may involve the second, rather than the first, coordination shell of the Mo atom. Given the high temperature reducing atmosphere under which catalytic reactions are generally performed using TMS, there is little doubt that the vacancies exist in significant amounts. Several structure–activity relationships strongly support this idea. The presence of optical defects, which absorb below the band gap, has been associated with sulfur-deficient molybdenum atoms and linearly correlates with the HDS of DBT. Another example, though performed over a supported catalyst, clearly indicates that the removal of sulfur is critical for the hydrogenation and isomerization of 1,3-pentadiene.[22] Finally, the well-known inhibition effect of H_2S is found in many reactions.

As described in the section on electronic structure above, the addition of a promoter also results in the modification of the sulfur species on the surface. A sulfur atom shared by a Co atom and a Mo atom is more likely to be removed than a sulfur shared by two Mo atoms. This is observed in the reactivity of small Co/Mo/S clusters.

5.6 Summary

A wide variety of bulk sulfide catalysts can be prepared by using conventional synthetic methods, which include low temperature precipitation and the sulfidation of oxides or salts. The characterization of bulk sulfides is often obtained through the use of a broad arsenal of techniques. Bulk characterization is generally achieved using X-ray diffraction and elemental analysis, whereas morphology and particle size, being more difficult to obtain, require such techniques as electron microscopy, EXAFS, and BET. Widely used for surface characterization—which is usually problematic because of its dependence on the particular system under investigation— are EXAFS, NMR, PDS, TEM, and chemisorption. When the structure is highly anisotropic (having layered sulfides), the characterization becomes very complex and challenging.

As supplies of clean petroleum feedstocks dwindle, industry will be called upon to process and upgrade "dirtier" feedstocks or find alternate routes for the production of transportation fuels and chemicals. Catalysts based on TMS materials are currently used for such processes, but will doubtless fall short of requirements for future use. Much progress has been made in understanding the properties of TMS-based catalysts, but still there remain basic problems that prevent further advances. A major impediment arises from the shortcomings of the characterization methods in analyzing highly anisotropic catalyst structures. Also, better control of the catalytic surface is needed. These challenges for the researcher will doubtless be resolved, and new applications and new TMS catalysts will be discovered in the future.

Bulk sulfide catalysts represent a unique class of heterogeneous catalysts, which possess an unusual resistance to poisoning. Because of this, they may be described as "tough catalysts for tough jobs."

References

1   O. Weisser and S. Landa. *Sulphide Catalysts, Their Properties and Applications.* Pergamon Press, New York, 1973.

2   R. R. Chianelli. *International Reviews in Physical Chemistry.* **2**, 127, 1982.

3   R. L. Seiver and R. R. Chianelli. U.S. Patent 4 430 443, 1984.

4   G. Hagenbach, P. Courty, and B. Delmon. *J. Catalysis.* **23**, 295, 1971.

5   R. Candia, B. S. Clausen, and H. Topsøe. *Bull. Soc. Chim. Belg.* **90**, 1225, 1981.

6   T. C. Ho, A. C. Young, A. J. Jacobson, and R. R. Chianelli. U.S. Patent 4 591 429, 1986.

7   T. R. Halbert, S. A. Cohen, and E. I. Stiefel. *Organometallics.* **4**, 1689, 1985.

8   M. Angulo, F. Mauge, J. C. Duchet, and J. C. Lavalley. *Bull. Soc. Chim. Belg.* **96**, 925, 1987.

9   V. A. Burmistrov, A. N. Startsev, and Y. I. Yermakov. *React. Kinet. Catal. Lett.* **24**, 365, 1984.

10   R. R. Chianelli, A. F. Ruppert, S. K. Behal, B. H. Kear, A. Wold, and R. Kershaw. *J. Catal.* **92**, 56, 1985.

11   T. R. Halbert, T. C. Ho, E. I. Stiefel, R. R. Chianelli, and M. Daage. *J. Catal.* **130**, 116, 1991.

12   D. J. Vaughan and J. R. Craig. *Mineral Chemistry of Metal Sulfides.* Cambridge University Press, 1978.

13   J. D. Passaretti, R. R. Chianelli, A. Wold, K. Dwight, and J. Covino. *J. Solid State Chem.* **64**, 365, 1986.

14   K. F. McCarty, J. W. Anderegg, and G. L. Schrader. *J. Catal.* **93**, 375, 1985.

15   O. Sørensen, B. S. Clausen, R. Candia, and H. Topsøe. *Appl. Catal.* **13**, 363, 1985.

16   C. B. Roxlo, M. Daage, A. F. Ruppert, R. R. Chianelli. *J. Catal.* **100**, 176, 1986.

17   R. Prins, V. H. J. De Beer, and G. A. Somorjai. *Catal. Rev.-Sci. Eng.* **31**, 1, 1989.

18   J. Bachelier, J. C. Duchet, and D. Cornet. *Bull. Soc. Chim. Belg.* **93**, 743, 1984.

19   N.-Y. Topsøe and H. Topsøe. *J. Catal.* **84**, 386, 1983.

20   Z. Shuxian, W. K. Hall, G. Ertl, and H. Knözinger. *J. Catal.* **100**, 167, 1987.

21   M. M. Disko, M. M. J. Treacy, S. B. Rice, R. R. Chianelli, J. B. Gland, T. R. Halbert, and A. F. Ruppert. *Ultramicroscopy.* **23**, 313, 1987.

22   S. Kasztelan, L. Jalowiecki, A. Wambeke, J. Grimblot, and J. P. Bonnelle. *Bull. Soc. Chim. Belg.* **96**, 1003, 1987.

23   R. R. Chianelli and M. Daage. *Studies in Surface Sci. and Catal.* **50**, 1, 1989.

24   M. J. Ledoux, O. Michaux, G. Agostini, and P. Panissod. *J. Catalysis.* **96**, 189, 1985.

25   H. Topsøe, B. S. Clausen, R. Candia, C. Wivel, and S. Mørup. *J. Catalysis.* **68**, 433, 1981.

26   B.S. Clausen, W. Niemann, P. Zeuthen, and H. Topsøe. *Preprints of the Division of Petroleum Chemistry.* **35**, 208, 1990.

27   S. P. A. Bouwers and R. Prins. *Preprints of the Division of Petroleum Chemistry.* **35**, 211, 1990.

28   M. J. Ledoux, Y. Segura, and P. Pannissod. *Preprints of the Division of Petroleum Chemistry.* **35**, 217, 1990.

29   T. A. Pecoraro and R. R. Chianelli. *J. Catal.* **67**, 430, 1981.

30   M. Ternan. *J. Catal.* **104**, 256, 1987.

31   S. Harris and R. R. Chianelli. *Chem. Phys. Lett.* **101**, 603, 1983.

32   R. R. Chianelli. *Catal. Rev.-Sci. Eng.* **26**, 361, 1984.

33   S. Harris and R. R. Chianelli. *J. Catal.* **98**, 17, 1986.

Supported Metal Sulfides

ROEL PRINS

Contents

6.1 Introduction

The major industrial application for supported metal sulfides is the removal of sulfur, nitrogen, oxygen, and metals from oil by reductive treatments in so-called hydrotreating processes. Of particular importance today, this purification process diminishes air-polluting emissions of sulfur and nitrogen oxides, which contribute to acid rain. Most oil streams in a refinery must be hydrotreated, since most catalysts used for the processing of oil products cannot tolerate sulfur and metals. Hydrotreating is therefore the largest application of industrial catalysis on the basis of the amount of material processed per year, and hydrotreating catalysts constitute the third largest catalyst business (after exhaust gas catalysts and fluid cracking catalysts).

Hydrotreating catalysts contain molybdenum and either cobalt or nickel supported on γ-Al_2O_3. A sulfided Co–Mo or Ni–Mo catalyst supported on Al_2O_3 has a much higher catalytic activity for the removal of S, N, and O atoms than a sulfided Mo/Al_2O_3 catalyst; and a sulfided Co/Al_2O_3 and Ni/Al_2O_3 catalyst has a much weaker catalytic activity. Based on this, molybdenum has always been considered to be the catalyst and cobalt and nickel the promoters of the Mo activity.[1]

Cobalt is mainly used as a promoter for sulfided Mo/Al_2O_3 in hydrodesulfurization (HDS), whereas nickel is favored in hydrodenitrogenation (HDN). Hydrotreating catalysts often contain modifier elements, such as P, B, F or Cl, that influence the catalytic and mechanical properties of the catalyst.

Hydrotreating catalysts originated in the 1920s with German researchers developing catalysts to liquefy coal, but the structure of these catalysts and the mechanism of their catalytic action was not understood until the 1970s. At that time, it was established that under actual catalytic conditions the majority of the molybdenum in industrial hydrotreating catalysts is present in the form of small MoS_2 particles in the pores of the γ-Al_2O_3 support. Although there was much discussion on the role and structure of cobalt and nickel promoters, it was not until the 1980s that the location of the promoter ions in the hydrotreating catalysts was known, and only recently have publications described the role of modifiers like phosphate and chlorine or fluorine.

This chapter deals mostly with Co–Mo–P/Al_2O_3 and Ni–Mo–P/Al_2O_3 hydrotreating catalysts, since they are used far more often than other supported metal sulfides. Tungsten may also be used in hydrotreating catalysts, but its industrial use is limited by its high cost. Similarities in the chemical properties of tungsten and molybdenum make the characterization of tungsten-based catalysts equivalent to that of molybdenum-based catalysts. Other transition metal sulfides have not yet been widely used in industrial applications. Uses for MoS_2-based catalysts other than hydrotreating, such as Fischer–Tropsch and water–gas shift catalysis, are still in the research stage.

In this chapter, we follow the synthesis of a hydrotreating catalyst through its impregnation, drying, calcination, and sulfidation. Also, characterization of the catalyst through spectroscopic and other techniques is used to determine which Mo, Ni (or Co), and P species are present in the impregnating solution, which species adsorb on the support surface, and how these species are transformed during calcination and sulfidation into the final catalytically active species. For the Mo-only catalyst, a more or less complete picture of its synthesis from impregnating solution to the final catalytically active species is described. For the Ni–Mo–P catalyst, the precise relationship between the properties of the impregnating solution and those of the final catalyst is not yet known. However, recent work by researchers has indicated that it is only a question of time before these relationships are understood, thus permitting the scientific design of hydrotreating catalysts.

6.2   Structure of the Oxidic Catalyst

Hydrotreating catalysts are industrially prepared by the pore volume impregnation of γ-Al_2O_3 with an aqueous solution of $(NH_4)_6Mo_7O_{24}$, $Co(NO_3)_2$, or $Ni(NO_3)_2$ and a phosphate containing salt, which is followed by drying, calcination, and sulfidation. The impregnation can be done in successive steps—with intermediate drying and calcination (heating in air)—or by a co-impregnation of all three inorganic materials, followed by drying and calcination. Phosphate is added during preparation as phosphoric acid or as ammonium phosphate to enhance the solubility of molybdate by the formation of phosphomolybdate complexes.

Infrared spectroscopy (IR), Raman spectroscopy, and NMR investigations have shown the types of complexes present in the impregnating solution.[2] In high pH solutions, MoO_4^{2-} complexes are most stable, and in low pH solutions, $Mo_7O_{24}^{6-}$.

$$Mo_7O_{24}^{6-} + 8OH^- \rightleftharpoons 7MoO_4^{2-} + 4H_2O \tag{6.1}$$

It is important to know in which form Mo is adsorbed on the Al_2O_3 surface. Al_2O_3 has its point of zero charge (PZC) at a pH of about 8, so its surface is positively charged at low pH and negatively charged at high pH. This means that the adsorption of negatively charged molybdate anions occurs best at low pH, when pH < PZC. At low pH, a molybdate solution contains mainly $Mo_7O_{24}^{6-}$ anions, but this does not necessarily mean that $Mo_7O_{24}^{6-}$ will be adsorbed on the Al_2O_3 surface at low pH. Since OH^- ions are released when heptamolybdate anions adsorb on the surface, the pH of the solution in the Al_2O_3 pores will increase. As a consequence, the $Mo_7O_{24}^{6-}$ ions will be depolymerized to MoO_4^{2-} ions, and because of their higher charge/radius ratio, the MoO_4^{2-} ions may be preferentially adsorbed. On the other hand, the OH^- ions released in the adsorption reaction shown in Equation 6.2 may be removed by the buffering action of Al_2O_3.

$$6[AlOH_2^+]OH^- + Mo_7O_{24}^{6-} \rightleftharpoons [AlOH_2^+]_6Mo_7O_{24}^{6-} + 6OH^- \tag{6.2}$$

The time differential observation of the perturbed angular correlation of γ-rays emitted from radioactive $^{99}Mo \rightarrow {}^{99}Tc$ makes it possible to distinguish different Mo species by their nuclear quadrupole interaction. Time differential perturbed angular correlation (TDPAC) measurements showed that heptamolybdate anions are preferentially adsorbed from $(NH_4)_6Mo_7O_{24}$ solutions at pH < PZC and at high Mo concentrations.[3] Thus, when an aqueous $(NH_4)_6Mo_7O_{24}$ solution (pH ≈ 5.5) is applied to a pore volume impregnation at a Mo concentration high enough to reach a monolayer of molybdate on the Al_2O_3 surface (about 6 at Mo/nm^2 Al_2O_3 BET surface area), mainly $Mo_7O_{24}^{6-}$ is adsorbed on the surface. Recent ^{95}Mo NMR measurements of the adsorbed species on the Al_2O_3 surface were in agreement with this conclusion.[4] The large line widths and small differences in chemical shifts among the different molybdate complexes prevented, unfortunately, a more detailed evaluation of the adsorbed state.

When molybdate as well as phosphate is present in solution, phosphomolybdate anions like $P_2Mo_5O_{23}^{6-}$, $PMo_9O_{31}(OH)_3^{6-}$, $PMo_{11}O_{39}^{7-}$, $PMo_{12}O_{40}^{3-}$, and $P_2Mo_{18}O_{62}^{6-}$ are present, depending on the pH and the P and Mo concentrations.[5] A ^{95}Mo NMR investigation[6] of a $(NH_4)_6Mo_7O_{24} + H_3PO_4$ solution (P:Mo = 2:5) in the pores of the alumina showed only the peak of MoO_4^{2-}, which was taken as proof for the decomposition of $P_2Mo_5O_{23}^{6-}$:

$$P_2Mo_5O_{23}^{6-} + 6[AlOH_2^+]OH^- \longrightarrow [AlOH_2^+]P_2Mo_5O_{23}^{6-} + 6OH^-$$

$$P_2Mo_5O_{23}^{6-} + 8OH^- \rightleftharpoons 5MoO_4^{2-} + 2HPO_4^{2-} + 3H_2O$$

Adsorption of $P_2Mo_5O_{23}^{6-}$ by ion exchange releases OH^- ions, which decompose $P_2Mo_5O_{23}^{6-}$ to molybdate and phosphate, both of which are strongly adsorbed on the alumina surface by ion exchange. This does not constitute proof, however, for a complete decomposition of the phosphomolybdate complex. The ^{95}Mo NMR signal of MoO_4^{2-} in the liquid state may equally as well have arisen from a minor decomposition of $P_2Mo_5O_{23}^{6-}$. Solid state ^{31}P and ^{95}Mo NMR measurements would be more appropriate for answering the question of whether phosphomolybdate completely decomposes on the alumina surface. In a study by van Veen et al.,[7] Raman and ^{31}P solid-state NMR spectra of alumina impregnated with $H_3PMo_{12}O_{40}$ indicated the presence of $P_2Mo_5O_{23}^{6-}$ and HPO_4^{2-} on the surface. Complete decomposition of $PMo_{12}O_{40}^{3-}$ by OH^- to $P_2Mo_5O_{23}^{6-}$ and $Mo_7O_{24}^{6-}$ and partial decomposition of $P_2Mo_5O_{23}^{6-}$ to HPO_4^{2-} and $Mo_7O_{24}^{6-}$ had apparently taken place. The incomplete decomposition of $P_2Mo_5O_{23}^{6-}$ was ascribed to an insufficient number of OH^- groups at the alumina surface. If this is true, decomposition of phosphomolybdates to molybdates and phosphate seems limited only by the number of OH^- groups present.

After drying, hydrotreating catalysts are usually calcined in air at a temperature of about 500 °C. After subsequent air exposure, they rapidly become rehydrated and contain hydroxyl groups as well as adsorbed water. The Raman spectrum of the calcined and air-exposed catalyst resembles that of the heptamolybdate complex, though the width of the band indicates distortion of the surface complex[2, 8] (see Figure 6.1). TDPAC revealed that the heptamolybdate may amount to 90% of the total Mo of the sample.[3] Raman studies of Mo/SiO_2 and Mo/Al_2O_3 catalysts[8] have shown that the structures formed by hydrated molybdenum oxide overlayers on oxidic supports are governed by the acid–base properties of the catalysts, especially by the point of zero surface charge. In situ Raman and EXAFS[9] measurements showed that the structure of calcined samples (which had not been exposed to air) differs from that of rehydrated samples. The presence of alumina OH groups on calcined Mo/Al_2O_3—as observed using IR and ^1H-NMR spectroscopy—even at high Mo loading, indicates that the heptamolybdate complexes do not cover the Al_2O_3 surface completely as a monolayer. The polymolybdate phase probably consists of small islands that can be several layers thick, leaving part of the alumina surface uncovered. When Mo/Al_2O_3 is calcined at higher temperatures ($\geq$ 650 °C), $Al_2(MoO_4)_3$ begins to form, as evidenced by the ^{27}Al-NMR line position. On the other hand, high-temperature calcination of $Mo–P/Al_2O_3$ catalysts first leads to $AlPO_4$ formation and, only above 700 °C, to $Al_2(MoO_4)_3$ formation.

There is substantial knowledge of the structure and composition of Mo/Al_2O_3 and $Mo-P/Al_2O_3$ catalysts during all preparation steps (impregnation, drying, calcination, and sulfiding), but the structure and composition of these catalysts in the

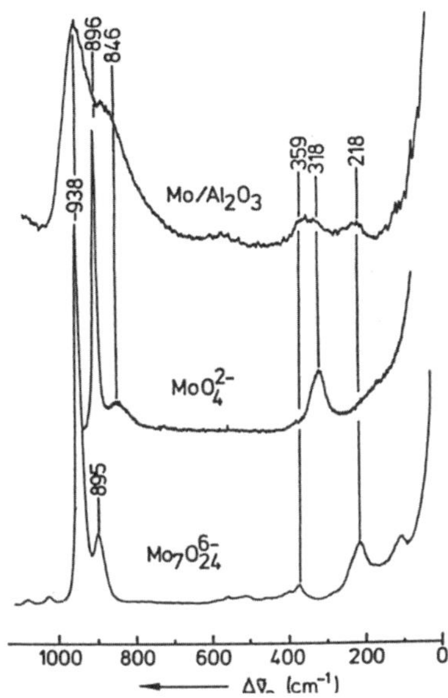

Figure 6.1   **Raman spectra of calcined and air-exposed Mo/Al₂O₃ and of molybdate monomer and heptamer in solution.**

presence of cobalt or nickel is less precise. It is known that Co and Ni cations may interact strongly with the alumina surface and that after calcination, a surface spinel with Ni^{2+} in octahedral and tetrahedral holes may be formed. At higher metal loading, NiO (or Co_3O_4) crystallites are formed on the support surface. This has been confirmed by magnetic susceptibility, UV-VIS, IR, XPS, Mössbauer, and X-ray absorption studies.[2]

When Mo and Ni (or Co) are put on the support surface, the Ni and Mo structure does not seem to be much different from that in a Mo-only or Ni-only catalyst. Raman and TDPAC studies[2] indicate the existence of polymolybdate complexes on the support surface in the promoted catalyst, and Mössbauer studies[10] show Co in tetrahedral and octahedral positions at low Co loading and Co_3O_4 at high loading. Thus, it seems that the Mo and promoter ions behave independently and are not in contact. On the other hand, the order of impregnation and calcination—Mo followed by Ni or vice versa—does play an important role in the activity of the final sulfided catalyst; catalysts in which the support was impregnated first with a Mo-containing solution invariably have a higher activity. Attempts have been made to explain this by describing interactions between the Mo and Ni (Co) ions on the support surface. For instance, it has been suggested that the Ni ions interact with the polymolybdate phase by forming a metal-heteropolymolybdate. This would allow the promoter ions to stay at the surface and be close to the Mo ions. During the subsequent sulfidation, the Ni and Mo ions would stay in close proximity, and the active Ni–Mo–S structure (which is discussed further below) can be more efficiently formed. However, there is no direct proof for the above Mo–Ni contact (as in Ni-heteropolymolybdate). Of course, proof for such a contact is difficult to give since the first coordination shell of Ni^{2+} in such a compound will be an octahedral oxygen surrounding, and thus indistinguishable from, Ni^{2+} in an octahedral hole in the alumina lattice.

Mössbauer spectroscopy of $Co–Mo/Al_2O_3$ catalysts shows that at higher calcination temperatures more and more Co^{2+} ions dissolve in the alumina.[10] As discussed below, this leads to less active catalysts, probably because these ions are lost for catalysis or because they are not able to form the catalytically active Co–Mo–S structure even if they diffuse back to the surface during sulfidation.

6.3   Structure of the Sulfidic Catalyst

Structure of Molybdenum

The oxidic catalyst precursors formed during the sequence of impregnation, drying, and calcination are transformed into the actual hydrotreating catalysts through sulfidation in a mixture of H_2 and a sulfur-containing molecule such as H_2S, thiophene, CS_2, dimethyldisulfide, or those that exist in an oil stream. In the laboratory, H_2S is a good choice. The properties of the final sulfidic catalyst depend strongly on the calcination and sulfidation steps. Calcination at high temperatures induces a strong interaction between the support and the Mo and Co (or Ni) ions in the oxidic state, making it difficult to transform the compounds formed into sulfides. The higher the calcination temperature, the higher the sulfidation temperature needed to reach the best hydrotreating activity, though at too high a temperature the metal oxides, as well as the metal sulfides, will sinter. Silvy et al.[11] have determined optimum calcination and sulfidation temperatures to be in the range of 400–500 °C. Sulfidation conditions have also been studied using temperature-programmed sulfidation,[12] in which the oxidic catalyst is heated in a flow of H_2S and H_2 and measurements are taken of the consumption of H_2S and H_2 and the evolution of H_2O. In this process, with the temperature slightly above 20 °C, H_2S is taken up and H_2O is given off, indicating a sulfur-oxygen exchange reaction. At higher temperatures, a reduction takes place, with concomitant H_2 consumption and H_2S evolution. At still higher temperatures, further sulfidation takes place (see Figure 6.2).

Because of its good textural and mechanical properties, γ-Al_2O_3 is the preferred choice as the support for industrial hydrotreating catalysts. Hydrotreating catalysts have to be regenerated several times during their life, putting high demands on the catalyst support. Regeneration consists in burning-off coke from the catalyst with air, purging the catalyst to avoid explosive conditions, and resulfiding the oxidized catalyst. At one time, catalyst regeneration was carried out on site in the reactor proper, but recently, off-site regeneration by specialized firms has become commonplace.

Throughout the impregnation, drying, and calcination steps of catalyst preparation, changes occur in the support texture that are not completely undone in the

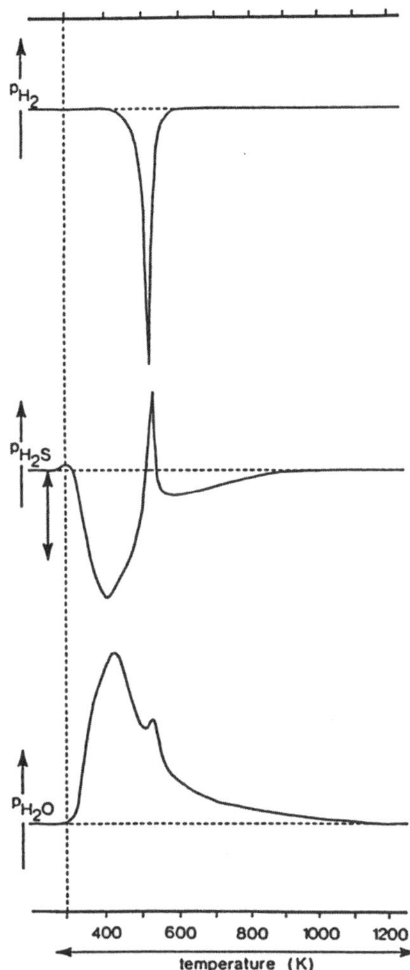

Figure 6.2 H$_2$, H$_2$S, and H$_2$O consumptions (negative) and evolutions (positive) in the temperature-programmed sulfidation of calcined Mo/Al$_2$O$_3$.[12]

sulfidation step. All catalyst components—Mo, Co (or Ni), and P—interact in some way with the support during the oxidic stages, but phosphate interacts especially strongly with the Al$_2$O$_3$, forming AlPO$_4$. As a result of phosphate interaction, micropores are closed, the surface area and pore volume decrease, and the mechanical and thermal stability of the resulting catalyst increase.[13] Reductive sulfidation does not influence the AlPO$_4$ structure, so the composition and texture of the support of a hydrotreating catalyst is different from that of the starting γ-Al$_2$O$_3$ material. The support consists of part γ-Al$_2$O$_3$ and part AlPO$_4$, and has a substantially different pore size distribution than does γ-Al$_2$O$_3$.

During sulfidation—as well as during actual hydrotreating—the conditions are highly reducing, with H$_2$S always present; thermodynamics predict that molybdenum be in the MoS$_2$ form. Indeed, extended X-ray absorption fine-structure

(EXAFS) studies[14] of Mo K-edge absorption spectra demonstrate that the average Mo ion in sulfided Mo/Al_2O_3 catalysts has the same environment as a Mo ion in MoS_2, the only difference being that the catalyst has an average of less than six molybdenum neighbors surrounding each Mo ion, which is the same as for pure MoS_2. Since EXAFS is a bulk technique used to determine the environment of both surface and interior Mo ions, these studies indicate that the percentage of surface Mo ions is substantial and that the MoS_2 particles on the support surface are small.

MoS_2 has a layered lattice, and the sulfur–sulfur interaction between sandwich domains of MoS_2 layers is weak. Crystals grow in the form of platelets, with a relatively large dimension parallel to the basal sulfur planes and a small dimension perpendicular to the basal plane. A high-resolution transmission electron microscopy (HR-TEM) study by Hayden and Dumesic[15] of hydrodesulfurization (HDS) model catalysts, which consist of MoS_2 crystallites on planar Al_2O_3, showed that MoS_2 crystallites occur in the form of platelets, with a height-to-width ratio between 0.4 and 0.7. Some of these platelets were oriented with their basal plane parallel to the Al_2O_3 surface, and some were oriented at a nonzero angle to the surface, suggesting that the MoS_2 platelets are bonded to the Al_2O_3 surface by Mo–O–Al bonds (see Figure 6.3).

Structure of Cobalt and Nickel

In the sulfidic form, cobalt may be present as Co_9S_8 crystallites on the support, as cobalt ions adsorbed on the edge surface of MoS_2 crystallites (the so-called Co–Mo–S phase, as discussed below), and in tetrahedral sites in the γ-Al_2O_3 lattice (see Figure 6.4). A similar situation holds for nickel, which can be present both as segregated Ni_3S_2, in the form of Ni–Mo–S, and as Ni ions in the support. A sulfided catalyst contains a relatively large amount of either Co_9S_8 or the Co–Mo–S phase, depending on the relative concentrations of cobalt and molybdenum in it and depending also on its pretreatment. Mössbauer experiments demonstrated that the structure of the catalyst in the sulfided state is predetermined by the structure of the oxidic precursor.[10] For example, Co_3O_4 transforms into Co_9S_8; cobalt ions in octahedral support sites transform into the Co–Mo–S phase; and cobalt ions in tetrahedral support sites remain in their positions (see Figures 6.5A and 6.5B). By combining Mössbauer studies with catalytic activity studies, Wivel et al.[16] established that the promoter effect of cobalt is related to the cobalt ions in the Co–Mo–S phase, not to separate Co_9S_8. Figure 6.5C demonstrates the strong promoter effect of Co on Mo, with small amounts of Co strongly increasing the thiophene

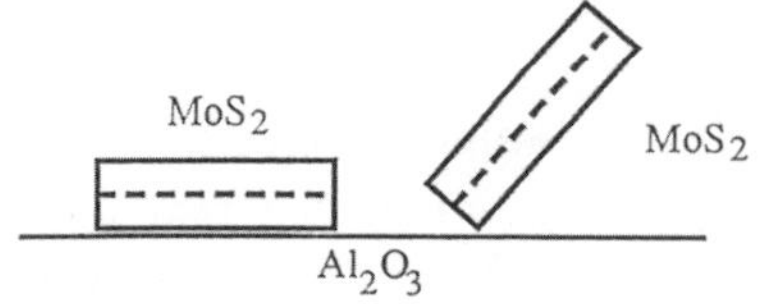

Figure 6.3 **Basal and edge bonding of MoS$_2$ on the Al$_2$O$_3$ support surface.**

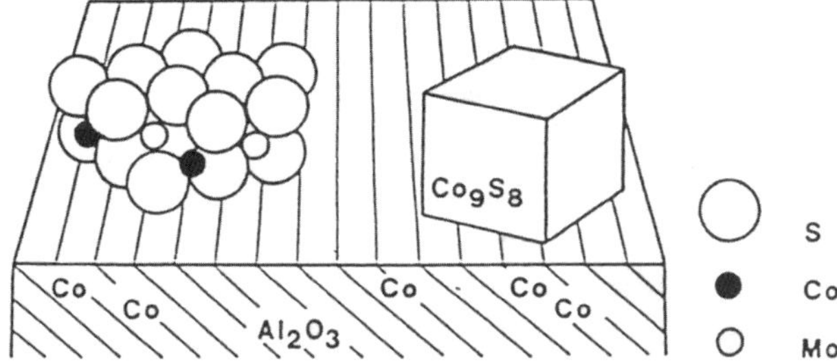

Figure 6.4 Three forms of Co in sulfided Co–Mo/Al$_2$O$_3$ catalysts: active sites at the MoS$_2$ edges (the Co–Mo–S phase); segregated Co$_9$S$_8$; and Co^{2+} ions in the support.[16]

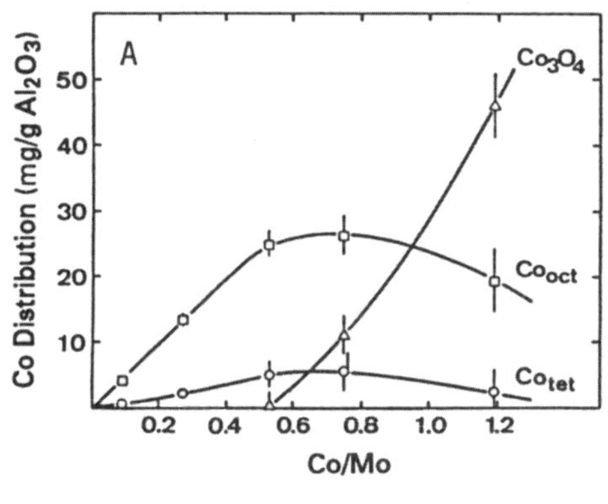

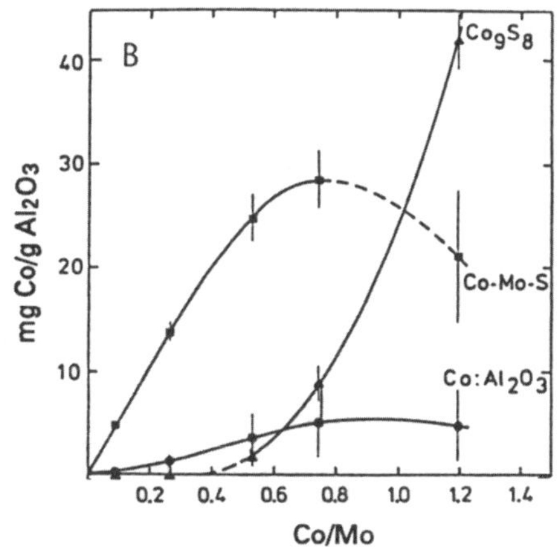

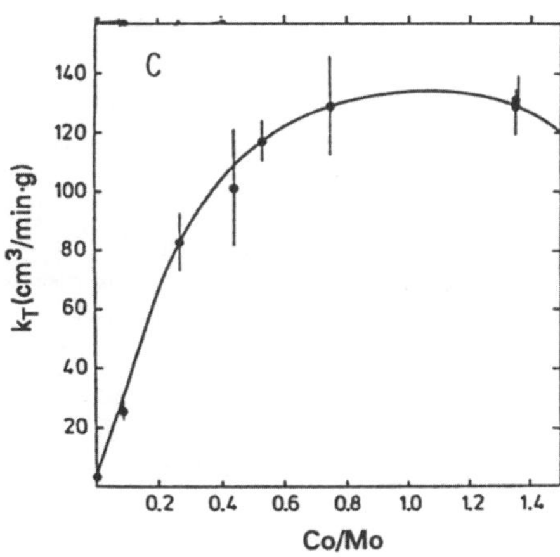

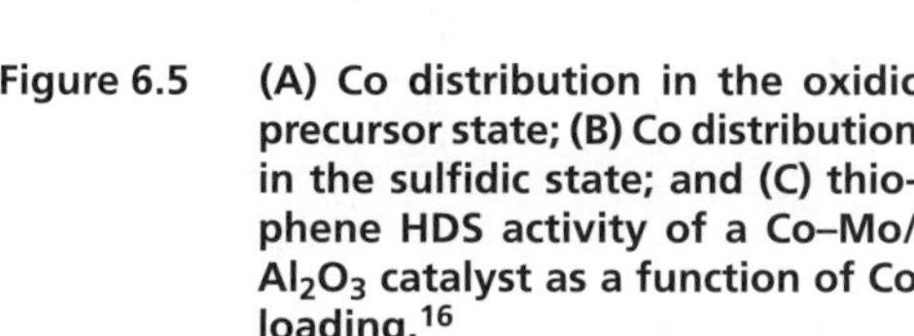

Figure 6.5 (A) Co distribution in the oxidic precursor state; (B) Co distribution in the sulfidic state; and (C) thiophene HDS activity of a Co–Mo/Al$_2$O$_3$ catalyst as a function of Co loading.[16]

HDS activity. Figure 6.5B indicates that initial amounts of Co lead preferentially to Co–Mo–S; Co_9S_8 is formed only at higher Co loadings, where the activity has already leveled off.

The adsorption of Co on MoS_2 (in the so-called edge decoration site shown in Figure 6.6) is thermodynamically unexpected for two reasons: (1) because the most stable phase of cobalt under sulfidic conditions is Co_9S_8 and (2) because solid-state chemistry studies show that $CoMo_2S_4$ is catalytically inactive and that cobalt and nickel do not form ternary compounds with MoS_2. Farragher and Cossee[17] suggested that cobalt ions at the MoS_2 edges are located between subsequent MoS_2 layers in a so-called surface-intercalation structure (see Figure 6.6). However, the Topsøe infrared study[18] of the adsorption of NO molecules on a series of sulfided Co–Mo/Al_2O_3 catalysts indicated that the edge decoration model is more likely. The IR spectrum of NO molecules adsorbed on Co ions is different from that of NO molecules on Mo ions. When the Co loading is increased at fixed Mo loading, the spectrum of NO adsorbed on Co sites is observed to increase in intensity, whereas the spectrum of NO adsorbed on Mo sites decreases in intensity. If the Co ions were in the intercalation location, the intensity of the NO on Co spectrum would have increased, as it in fact did, but the intensity of the NO on Mo spectrum would have stayed constant. However, cobalt ions in the edge decoration location cover Mo ions and block adsorption of NO on these Mo ions. Therefore, the observed behavior is in accordance with the edge decoration location. The number of edge and corner Mo and promoter sites have been calculated as a function of MoS_2 particle size and compared with experimental results.[19] The reasonable fit between the predictions and experimental results indicates that the assumptions underlying the edge decoration model are realistic.

EXAFS studies have proven that the promoter ions are indeed linked to the MoS_2 edges, as predicted in the edge decoration model. The study by Bouwens et al.[20, 21] shows that in sulfided Ni–Mo catalysts supported on γ-Al_2O_3 (as well as on carbon), the Ni atoms are surrounded by five S atoms at 2.22 Å, by one or two Mo atoms at 2.8 Å, and by one Ni atom at 3.2 Å. These data are fully consistent with a model in which the Ni ions are located in the Mo plane at the MoS_2 edges in a square pyramidal coordination. The Ni ions are connected to the MoS_2 by four sulfur atoms with the fifth sulfur atom in the apical position in front of the Ni ion. A neighboring Ni atom is located at the next edge position at 3.2 Å (see Figure 6.7).

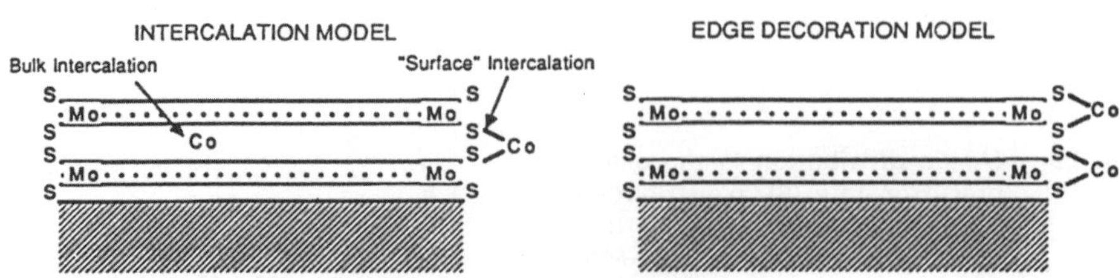

Figure 6.6   Models for the structure of the promoter of MoS$_2$.

Candia et al.[22] have claimed that there are two forms of the Co–Mo–S structure, with Co ions decorating the MoS_2 edges. A comparison of thiophene HDS activity data with the number of Co atoms in the Co–Mo–S structure obtained from ^{57}Co Mössbauer emission data shows that $Co–Mo/Al_2O_3$ catalysts sulfided at high temperature are intrinsically more active per Co ion in the Co–Mo–S phase than those sulfided at 400–500 °C. The more active phase was called Co–Mo–S II, and the less active phase Co–Mo–S I. The type I structure is said to be bonded to the support via Mo–O–Al linkages, whereas type II might have few, if any, of such linkages. The high activity of carbon-supported Co–Mo catalysts is ascribed to the presence of Co–Mo–S II because of a weak catalyst-carbon interaction,[23] and the high activity of $Co–Mo/Al_2O_3$ catalysts prepared in the presence of complexing agents like EDTA (ethylenediaminetetraacetic acid) and NTA (nitrilotriacetic acid) might be caused by the prevention of catalyst–support interactions, leading to the preferential formation of the type II structure.[24] Browne et al.[25] have recently observed that the passivation and resulfidation of Co–Mo and Ni–Mo catalysts induces a substantial increase in catalytic activity, the cause of which they postulate is a breaking of catalyst–support interactions and, after resulfidation, the formation of the type II phase. Since the Mössbauer and EXAFS signals of the type I and II structures do not differ at all, the local Co and Ni environments must be exactly the same for both types. This suggests that type I is less active than type II for steric reasons. Catalyst-support linkages in type I probably hinder the approach of reactant molecules to the catalytic sites.

6.4 Specific Surface Area

There are several methods for determining the (specific) surface area of supported metal catalysts, but no generally accepted method for supported metal sulfides. Metal sulfide catalysts are complex, consisting of several components. Even if it were possible to measure the dispersion of MoS_2 crystallites on the support, the dispersion of the Co or Ni promoter ions would still need to be determined. Furthermore, it is not sufficient to measure the *average* MoS_2 dispersion (surface to volume ratio), since only the edge surface, not the basal planes, is catalytically

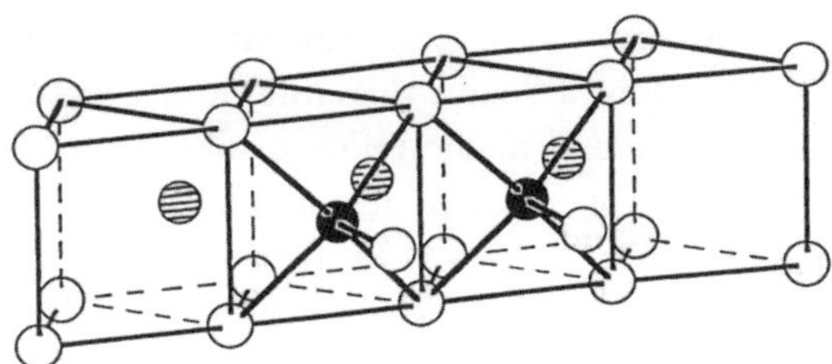

Figure 6.7   **Structure of the Ni atoms in the Ni–Mo–S phase at the MoS_2 edge (filled circles indicate Ni, empty circles S, and shaded circles Mo).**

active. Not surprisingly, in a study by Tauster et al.,[26] the total surface area of unsupported MoS_2 did not correlate with catalytic activity, but the edge surface area measured by means of O_2 chemisorption did. In O_2 chemisorption, the oxygen molecule is supposed to adsorb on a Mo cation and oxidize it; thus, in principle, it could be a valuable tool for determining the number of exposed Mo ions. But, in practice, the method has a limited applicability, because oxygen chemisorption does not stop at the oxidation of exposed Mo ions; it induces a surface reconstruction, and goes on to oxidize subsurface and deeper lying Mo ions. However, since O_2 chemisorption and HDS activity are both linearly dependent on crystal edge surface area, there is a certain linear relationship between O_2 chemisorption and HDS activity for a particular catalyst, with the linear rate depending on the class of catalyst (e.g., the linear relationship for MoS_2 catalysts differs from that for promoted MoS_2 catalysts). This relationship does depend on factors such as O_2 pressure, temperature, and exposure time during chemisorption and on the type of crystal surface (e.g., promoted or unpromoted). Since O_2 chemisorption reflects the general state of dispersion of the catalyst rather than that for specific sites, O_2 chemisorption cannot be used for the quantitative determination of active sites.[27]

CO and NO have also been tried as adsorbents for determining the active surface area of sulfide catalysts. Bachelier et al.[28] have proposed CO adsorption at O °C as a measure for the sulfide edge surface area; however, CO binds weakly to MoS_2 and only somewhat weaker to Al_2O_3. Infrared and adsorption studies[18] demonstrate a good correlation between NO chemisorption and HDS activity for MoS_2/Al_2O_3 catalysts, but not for promoted catalysts. XPS studies,[29] on the other hand, indicate that NO adsorbs dissociatively above 130 K on polycrystalline MoS_2 and that the resulting oxygen atoms lead to surface oxidation above 200 K. It is clear that a reliable chemisorption method for determining the specific surface area of supported metal sulfides has yet to be developed. Such determinations can be made using high-resolution transmission electron microscopy (HR-TEM). Unfortunately, this method is not available in every laboratory.

6.5   Structure–Reactivity Relationships

The removal of S, N, and O atoms from saturated molecules, such as mercaptans, thioethers, alkylamines, alcohols, and ethers, is relatively easy and occurs most probably through acid–base catalyzed elimination reactions. The removal of heteroatoms from aromatic ring structures is much more difficult; therefore, much attention has been devoted to the HDS of thiophene and to the HDN of pyridine and quinoline. The mechanism of the HDS of thiophene is presented in Figure 6.8. Two possible pathways are indicated: (1) thiophene is split with the aid of H_2 into butadiene and H_2S (the so-called hydrogenolysis step) and (2) thiophene is hydrogenated to tetrahydrothiophene, which then reacts to butadiene and H_2S (probably by a double elimination reaction). In both cases, the butadiene is quickly hydrogenated to butene, which isomerizes and is then further hydrogenated to butane.

                              SUPPORTED METAL SULFIDES   Chapter 6

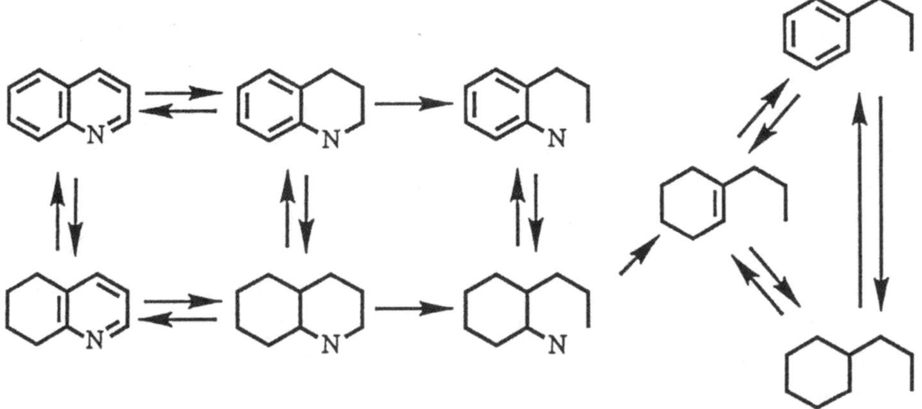

Figure 6.8   Reaction network of the HDN of quinoline to C_9 hydrocarbons and NH_3.

At high H_2 pressures, the dominant pathway for this reaction is via hydrogenation.[30] The pathway at 1 atm is currently being debated.

An example of an HDN reaction mechanism, for the N removal from quinoline, is presented in Figure 6.9. In this diagram, C–N bond breaking occurs only after hydrogenation of both aromatic rings (since the C–N bond strength in pyridine and aniline is much stronger than that in piperidine). As a consequence, hydrogen consumption is substantial in HDN. Process conditions are more severe in HDN than for HDS; they are a compromise between kinetics (high temperature) and thermodynamics (ring hydrogenation is favored at low temperatures and high H_2 pressures). Both hydrogenolysis and βH-elimination may be responsible for the final C–N bond breaking.

The following questions now arise: Which catalytic sites are responsible for the reaction steps in HDS and HDN? If Mo sites are the catalytically active sites, how are they promoted by Co and Ni, and what role does phosphate play?

Role of Molybdenum

It is often assumed that the catalytically active sites in a hydrotreating catalyst are Mo cations at the surface of the MoS_2 crystallites that have at least one sulfur

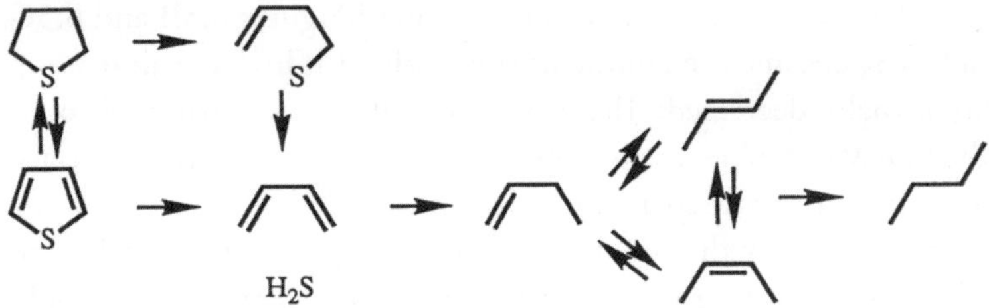

Figure 6.9   Reaction network of the HDS of thiophene to C_4 hydrocarbons and H_2S.

vacancy, which allows the reacting molecule to chemically bind to the Mo cation. Since sulfur anions in the basal planes of MoS_2 are more difficult to remove than anions at edges and corners, there will be a greater concentration of exposed Mo ions at the edges and corners, which is where catalysis will most likely occur. This has been confirmed by Farias et al.[31] In their study, an MoS_2 single crystal with a large ratio of basal plane over edge surface area had a low HDS activity, but the crystal's activity increased after the sputtering of sulfur atoms from its basal plane and the exposure of the Mo ions.

Desulfurization begins with adsorption of thiophene to the catalyst by π coordination of the whole thiophene ring to a Mo ion. Thiophene adsorption is followed by hydrogen transfer and the elimination of sulfur to form butadiene. The sulfur atom is subsequently removed from the catalyst surface by hydrogenation and H_2S desorption.

Role of the Promoter

The HDS (and HDN) activity of an MoS_2/Al_2O_3 catalyst increases substantially with the addition of a Co or Ni promoter. The promoter effect results in an increase in the number of Mo sites at the catalyst surface. Several explanations for the promoter function of Co and Ni have been proposed,[1] one theory being that the promoter creates more active sites by altering the texture of the catalyst surface, without itself being involved in the catalysis. The role of Ni could be to create more catalytically active Mo^{3+} sites,[32] or the promoter atoms could induce a surface reconstruction of the edges of the MoS_2 layers, providing more exposed Mo cations and, thus, an enhanced activity.[17] However, IR[18] and EXAFS[14] investigations have revealed that newly exposed Mo ions become covered by the sulfur and the promoter ions, indicating that if only Mo ions were active, the catalytic activity should decrease with the addition of Co or Ni ions.

In a model proposed by Delmon,[33, 34] segregated cobalt sulfide is the promoter, providing hydrogen atoms to MoS_2. These spilled-over H atoms create reduced centers on the MoS_2 surface, which become catalytically active sites. The Co_9S_8 can be said to have "remote control" over the MoS_2 surface. However, combined Mössbauer and activity studies[16] demonstrate that even when the presence of minor amounts of Co_9S_8 is excluded (at Co/Mo < 0.4), the catalytic activity still increased strongly with the addition of cobalt, as follows from a comparison of Figures 6.5B and 6.5C. Furthermore, when Co_9S_8 became the dominant cobalt phase at high cobalt loading, the catalytic activity actually decreased. The promotion effect was attributed to the cobalt present in the Co–Mo–S phase, with cobalt ions located at the MoS_2 surface; and a significant contribution from separate Co_9S_8 was excluded.

Catalysis researchers repeatedly discuss the idea that the promoter ion influences a neighboring Mo site, thus creating a much more active site. Particularly popular are those models in which electrons are donated from the promoter to the Mo ion, such a donation leading to a weakening of the Mo–S bond. Quantitative SCF-Xα

scattered wave molecular orbital calculations[35, 36] of $(MS_6)^{n-}$ complexes, where M is a transition metal ion, indicate that differences in catalytic activity among the various metal sulfides might be due to electronic factors, such as the electron occupancy of the highest occupied molecular orbital (HOMO) and covalency of the d-orbitals. Calculations on $(S_3MS_3MoS_3)^{m-}$ complexes, in which M was varied from V to Zn and in which an octahedral sulfur coordination around M and Mo was assumed, indicate that the promotion effect could be related to electron donation from promoter ion to Mo^{4+}. The energy levels of the e_g HOMOs of the Co and Ni complexes are especially suited for electron donation to the t_{2g} lowest unoccupied molecular orbital (LUMO) of Mo^{4+}.

Cobalt and nickel sulfide supported on carbon have a higher HDS activity than MoS_2/C. Duchet et al.[37] have suggested that the cobalt and nickel ions in the Co–Mo–S and Ni–Mo–S phases, respectively, might be the catalyst instead of the promoter. In the past, the idea of Co and Ni being the catalyst in sulfided Co–Mo and Ni–Mo systems had been rejected, because sulfided Co/Al_2O_3 and Ni/Al_2O_3 catalysts have a very low HDS activity. During the usual pretreatment of Co/Al_2O_3 and Ni/Al_2O_3 catalysts, cobalt and nickel ions interact strongly with Al_2O_3 and the low HDS activity is due to either (1) the metal ions not becoming sulfided at all during the sulfidation part of the process and, hence, not contributing to HDS activity or (2) the application of severe sulfidation conditions lowering the dispersion and activity of the metal ions. Carbon-supported cobalt and nickel sulfide catalysts demonstrate that when these catalysts are carefully prepared they can indeed have a high activity. The activity per Co atom in a sulfided Co–Mo/C catalyst compares much better with the activity per estimated surface Co atom in a sulfided Co/C catalyst than with the activity per estimated edge Mo atom in a MoS_2/C catalyst.[38] Further evidence for Co and Ni, rather than Mo, being the catalytic sites comes from the observation that the hydrogenation pattern of Co–Mo and Ni–Mo catalysts resembles that of sulfided Co and Ni catalysts, respectively, and is different from that of supported MoS_2. Mo EXAFS investigations[14, 20, 21] show that the number of sulfur ions coordinated around each Mo ion is about five in unpromoted Mo catalysts and very close to six in Co–Mo and Ni–Mo catalysts. This indicates that Mo is fully coordinated and cannot be catalytically active. Furthermore, a recent EXAFS study[39] of the Ni and Mo edges of a sulfided Ni–Mo/Al_2O_3 catalyst shows that selenophene (the Se analogue of thiophene) adsorbs on the catalyst through coordination to Ni and not to Mo.

Role of Phosphate

Phosphate enhances the solubility of molybdate in the impregnation solution and improves the mechanical and thermal stability of the support through $AlPO_4$ formation.[40] The $AlPO_4$ formation weakens the adsorption of molybdate and decreases the surface area of the support, resulting in diminished MoS_2 dispersion. On the other hand, the weaker interaction between molybdate and the phosphated

Al_2O_3 support may increase the sulfidability of Mo and induce the formation of type II Co–Mo–S and Ni–Mo–S structures, producing an enhanced activity. It may also lead to higher stackings than occur on pure Al_2O_3. HR-TEM shows that phosphate does indeed contribute to the stacking of the MoS_2 slabs, though it is not clear at present whether stacking leads to a change in activity. Successive MoS_2 slabs in a stack are bonded through weak van der Waals interactions, which will only have a minor electronic effect on the catalytic activity. The number of edge Mo ions in a pile of MoS_2 slabs is equal to that of the same number of equally sized single-slabs spread over the support surface (in an unstacked arrangement). The combined influence of dispersion, type I or type II, and stacking on hydrotreating activity is unknown at present, but the growing number of publications dealing with phosphate suggests that an explanation will soon be forthcoming.

Eijbouts et al.[42] have observed phosphate to have a minor effect on the HDS of thiophene over sulfided Ni–Mo/A_2O_3 catalysts, but a substantial effect on the HDN of quinoline. In their study, quinoline conversion increased by a factor of about two, the activation energy changed, and the saturates/unsaturates product ratio changed. Thus, the effect of phosphate cannot be due to a mere improvement in the dispersion of the Ni–Mo–S phase or to a change in the proportion of type I and type II sites, since both HDN and HDS would then have improved. The HDN and HDS have complex reaction networks, in which the hydrogenation reactions take place on Ni, Ni–Mo–S, or Mo sites and the C–S and C–N bond breaking reactions take place on Brönsted acid sites, such reactions being brought about by H_2S adsorption on the metal sulfide or the support or from the phosphate promoter. A hydrotreating catalyst is thus a bifunctional catalyst. In the HDN of quinoline, both the N-removal from decahydroquinoline and the preceding hydrogenations are slow,[43] thus precluding a rate-determining step treatment. But, in the thiophene HDS reaction network, the first step, which is assumed to take place on a Mo, Ni, or Ni–Mo–S site, is rate determining. The fact that phosphate has almost no effect on the HDS, but a substantial effect on the HDN of quinoline, indicates a dependency on the rate of the C–N or C–S bond breaking steps relative to that for the hydrogenation steps. Therefore, in the HDN of quinoline (with its slow N-removal from decahydroquinoline) phosphate has a beneficial effect; in the HDS of thiophene it has no effect; and in the HDN of pyrrole, in which the first hydrogenation step is rate determining, it should have no effect. Because of its positive effect on the HDN of pyridine-type molecules, and its predicted negative effect on that of pyrrole-type molecules, the effect of phosphate on real oil feeds may depend on the type of oil.

6.6 Summary

Modern spectroscopic techniques are important for the elucidation of the structure of Ni–Mo–P/Al_2O_3 and Co–Mo-P/Al_2O_3 hydrotreating catalysts. The Mössbauer method is invaluable in the simultaneous quantitative determination of the three forms in which Co can be present on the catalyst, providing the first clue to the catalytically active species. It is also very important for establishing a relationship between the three sulfidic and the three oxidic Co forms. Knowledge of these relationships contribute to better catalyst preparation. Unfortunately, Mössbauer equipment is not generally available, certainly not for quality control. Since the octahedral form of Co and Ni leads to the Co–Mo–S and Ni–Mo–S structure, the UV-VIS method may be used to optimize the octahedral Co or Ni coordination.

TDPAC spectroscopy determines which Mo complexes are adsorbed during impregnation and which are present after calcination, and EXAFS provides the proof for the existence of Co–Mo–S and Ni–Mo–S structures. Classic techniques, such as IR, Raman, and XPS spectroscopy also provide useful information. Solid-state NMR will become an important tool for studying hydrotreating catalysts when the line-width problem in[95] Mo NMR is overcome. Non-spectroscopic techniques, such as the TPS, TPR, and TPO thermal methods, provide valuable information as well.

The determination of catalyst dispersion still constitutes a problem—only high-resolution electron microscopy can at present be used with confidence for these measurements. A routine dispersion method, like the chemisorption technique, is not yet available. It should be realized that no technique on its own can provide answers to all questions that need be asked about the structure of a supported metal sulfide. It is through a combination of techniques that progress can be made.

References

1   R. Prins, V. H. J. de Beer, and G. A. Somorjai. *Catal. Rev.-Sci. Eng.* **31**, 1, 1989.

2   H. Knözinger. In *Proc. 9th Int. Congress Catal.* (M. J. Phillips and M. Ternan, Eds.) The Chemical Institute of Canada, Ottawa, 1988, p. 20.

3   T. Butz, C. Vogdt, A. Lerf, and H. Knözinger. *J. Catal.* **116**, 31, 1989.

4   J. C. Edwards, R. D. Adams, and P. E. Ellis. *J. Am. Chem. Soc.* **112**, 8349, 1990.

5   J. A. R. van Veen, O. Sudmeijer, C. A. Emeis, and H. de Wit. *J. Chem. Soc. Dalton Trans.* 1825, 1986.

6   W.-C. Cheng and N. P. Luthra. *J. Catal.* **109**, 163, 1988.

7   J. A. R. van Veen, P. A. J. M. Hendriks, E. J. G. M. Romers, R. R. Andrea, and A. E. Wilson. *J. Phys. Chem.* **94**, 5275, 5282, 1990.

8   C. C. Williams, J. G. Ekerdt, J.-M. Jehng, F. D. Hardcastle, A. M. Turek, and I. E. Wachs. *J. Phys. Chem.* **95**, 8781 and 8791, 1991.

9   M. de Boer, A. J. van Dillen, D. C. Koningsberger, J. W. Geus, M. A. Vuurman, and I. E. Wachs. *Catal. Lett.* **11**, 227, 1991.

10 C. Wivel, B. S. Clausen, R. Candia, S. Mørup, and H. Topsøe. *J. Catal.* **87**, 497, 1984.

11 R. Prada Silvy, P. Grange, and B. Delmon. *Stud. Surf. Sci. Catal.* **52**, 233, 1989.

12 P. Arnoldy, J. A. M. van den Heijkant, G. D. de Bok, and J. A. Moulijn. *J. Catal.* **92**, 35, 1985.

13 R. Lopez Cordero, N. Esquivel, J. Lazaro, J. L. G. Fierro, and A. Lopez Agudo. *Appl. Catal.*, **48**, 341, 353, 1989.

14 S. M. A. M. Bouwens, R. Prins, V. H. J. de Beer, and D. C. Koningsberger. *J. Phys. Chem.* **94**, 3711, 1990.

15 T. F. Hayden and J. A. Dumesic. *J. Catal.* **103**, 366, 1987.

16 C. Wivel, R. Candia, B. S. Clausen, S. Mørup, and H. Topsøe, *J. Catal.*, **68**, 453, 1981.

17 A. L. Farragher and P. Cossee. In *Proc. 5th Int. Congress Catalysis.* North-Holland, Amsterdam, 1973, p. 1301.

18 N. Y. Topsøe and H. Topsøe. *J. Catal.* **84**, 386, 1983.

19 S. Kasztelan, H. Toulhoat, J. Grimblot, and J. P. Bonnelle. *Appl. Catal.* **13**, 127, 1984.

20 S. M. A. M. Bouwens, D. C. Koningsberger, V. H. J. de Beer, S. P. A. Louwers, and R. Prins. *Catal. Letters.* **5**, 273, 1990.

21 S. P. A. Louwers and R. Prins. Forthcoming in *J. Catal.*

22 R. Candia, O. Sørensen, J. Villadsen, N. Y. Topsøe, B. S. Clausen, and H. Topsøe. *Bull. Soc. Chim. Belg.* **93**, 763, 1984.

23 J. P. R. Vissers, B. Scheffer, V. H. J. de Beer, J. A. Moulijn, and R. Prins. *J. Catal.* **105**, 277, 1987.

24 J. A. R. van Veen, E. Gerkema, A. M. van der Kraan, and A. Knoester. *J. Chem. Soc. Chem. Commun.* 1684, 1987.

25 V. M. Browne, S. Louwers, and R. Prins. *Catal. Today.* **10**, 345, 1991.

26 S. J. Tauster, T. A. Pecoraro, and R. R. Chianelli. *J. Catal.* **63**, 515, 1980.

27 W. Zmierczak, G. Murali Dhar, and F. E. Massoth. *J. Catal.* **77**, 432, 1982.

28 J. Bachelier, M. J. Tilliette, M. Cornac, J. C. Duchet, J. C. Lavalley, and D. Cornet. *Bull. Soc. Chim. Belg.* **93**, 743, 1984.

29 Z. Shuxian, W. K. Hall, G. Ertl, and H. Knözinger. *J. Catal.* **100**, 167, 1986.

30 H. Schulz, M. Schon, and N. M. Rahman. *Stud. Surf. Sci. Catal.* **27**, 201, 1986.

31 M. H. Farias, A. J. Gellman, G. A. Somorjai, R. R. Chianelli, and K. S. Liang. *Surf. Sci.* **140**, 181, 1984.

32 R. J. H. Voorhoeve and J. C. M. Stuiver. *J. Catal.* **23**, 228, 243, 1971.

33 B. Delmon. *Bull. Soc. Chim. Belg.* **88**, 979, 1979.

34 M. Karroua, P. Grange, and B. Delmon. *Appl. Catal.* **50**, L5, 1989.

35 S. Harris and R. R. Chianelli. *J. Catal.* **86**, 400, 1984.

36 S. Harris and R. R. Chianelli. *J. Catal.* **98**, 17, 1986.

37 J. C. Duchet, E. M. van Oers, V. H. J. de Beer, and R. Prins. *J. Catal.* **80**, 386, 1983.

38 J. P. R. Vissers, V. H. J. de Beer, and R. Prins. *J. Chem. Soc. Faraday Trans., Vol 1.* **83**, 2145, 1987.

39 A. N. Startsev, S. A. Shkuropat, V. V. Kriventsov, D. I. Kochubey, and K. I. Zamaraev. *J. Chem. Soc. Chem. Commun.* 6, 1991.

40 E. C. DeCanio, J. C. Edwards, T. R. Scalzo, D. A. Storm, and J. W. Bruno. *J. Catal.* **132**, 498, 1991.

41 R. C. Ryan, R. A. Kemp, J. A. Smegal, D. R. Denley, and G. A. Spinnler. *Stud. Surf. Sci. Catal.* **50**, 21, 1989.

42 S. Eijsbouts, J. van Gestel, J. A. R. van Veen, V. H. J. de Beer, and R. Prins. *J. Catal.* **131**, 412, 1991.

43 S. H. Yang and C. N. Satterfield. *Ind. Eng. Chem. Proc. Des. Dev.* **23**, 20, 1984.

Zeolites and Molecular Sieves

MARK E. DAVIS AND JOHN B. HIGGINS

Contents

7.1 Introduction

Due to their unique properties, zeolites and other molecular sieves are finding widespread use as adsorbents, ion exchangers, catalysts, and catalytic supports. The field of molecular sieve science is burgeoning at an extremely rapid pace. In 1970 alone, approximately 200 patents and 750 publications were issued that dealt with zeolites, whereas by 1985 the numbers had risen to approximately 700 patents and 2000 publications per year.[1] This trend should continue into the future, especially since the use of molecular sieves is now filtering into other areas of materials science such as ceramics, electronic materials, and drug release agents. In this chapter we define zeolites and molecular sieves and describe several widely available characterization techniques for the solution of their structure.

7.2 Structure of Zeolites and Molecular Sieves

In 1756, the Swedish mineralogist A. F. Cronstedt heated an unidentified silicate mineral and noticed that it readily fused, with marked intumescence, in the blowpipe flame. This observation led him to coin the term zeolite [derived from the Greek words *zeo* (to boil) and *lithos* (stone)] for minerals which behaved in this fashion. Since then, approximately 40 mineral, or natural, zeolites have been discovered. There are also over 30 synthetic zeolites and molecular sieves that do not have natural, structural analogs.[2]

Zeolites are crystalline aluminosilicates constructed from TO_4 tetrahedra (where T is a tetrahedral atom such as Si or Al), with each apical oxygen atom shared with an adjacent tetrahedron. Thus, the ratio O/T is equal to two. When tetrahedra of silicon and aluminum (refer to Figure 7.1) are connected to form an aluminosilicate structure, there is a negative charge associated with each aluminum atom balanced by a positive ion (M^+ in Figure 7.1) to give the structure electrical neutrality. Typical cations in natural zeolites are alkali metals (such as Na^+ and K^+) and alkaline earth (such as Ca^{2+} and Ba^{2+}); synthetic zeolites may contain both inorganic and organic cations, including quaternary ammonium ions and protons, H^+. Tectoaluminosilicates do not generally have Si/Al ratios of less than one, implying that an aluminum atom does not have another aluminum atom in its second coordination (the first coordination is always oxygen). Ratios of Si/Al greater than one are most commonly observed.

Figure 7.2 illustrates the construction of a common structural unit, the sodalite cage, with its constituent silicon, aluminum, and oxygen atoms. Notice that variations in the arrangement of the cages can produce a variety of zeolite structures. Not all zeolites contain cages. Zeolites can be comprised of channels (ZSM-5 in Figure 7.3), cages (NaX in Figure 7.3), and combinations of both. Zeolite void spaces range in size from approximately 4 to 13 Å. Meier and Olson[2] provide a

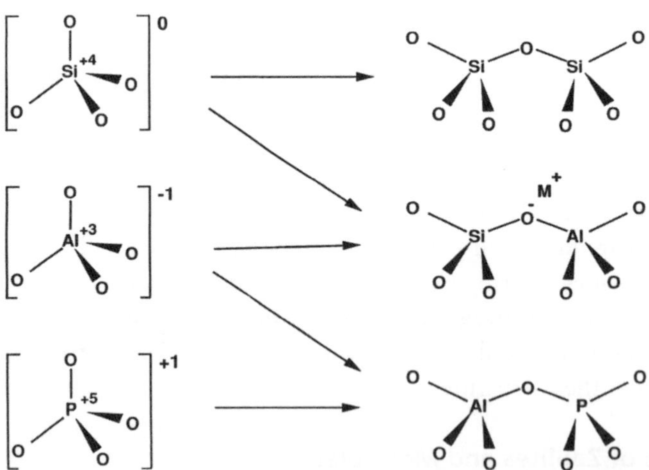

Figure 7.1   **TO_2 units in zeolites and aluminophosphates. (From M. E. Davis, *Ind. Eng. Chem. Res.*, 30, 1675, 1991. Reprinted by permission of the American Chemical Society.)**

complete list of the various structures. Zeolites are unique in that their pores are uniform in size and are in the same size range as small molecules (see Figure 7.3). They act as molecular sieves, discriminating among molecules on the basis of size: molecules smaller than the pore size are admitted into the crystal interior while those larger are not. These well-defined void spaces in a zeolite can be used to host a variety of interesting guest materials such as semiconductor clusters (the "quantum confined solids"), polymers, and other molecular-sized materials.

Strictly speaking, a zeolite is an aluminosilicate. Molecular sieves with framework atoms other than silicon and aluminum are not termed zeolites. The substitution of ions other than Al^{3+} and Si^{4+} (Reference 3 provides a fairly comprehensive list that includes iron, boron, chromium, beryllium, gallium, germanium, cobalt, vanadium, zinc, and titanium) into pure-silica or aluminosilicate frameworks produces molecular sieves normally called metallosilicates. Molecular sieves that do not contain silicon also exist. If tetrahedra containing aluminum and phosphorus are connected (see Figure 7.1) in a strict Al/P = 1, a neutral aluminophosphate or $AlPO_4$ framework is obtained. The designation $AlPO_4$-n has been used, the number n denoting a particular structure type (see Figure 7.3 for examples). Additional elements can be incorporated into the aluminophosphate based molecular sieves; silicon, magnesium, iron, titanium, cobalt, vanadium, zinc, manganese, gallium, germanium, beryllium, and boron have been combined with the $AlPO_4$'s to create a vast number of element-substituted aluminophosphate-based molecular sieves.[4]

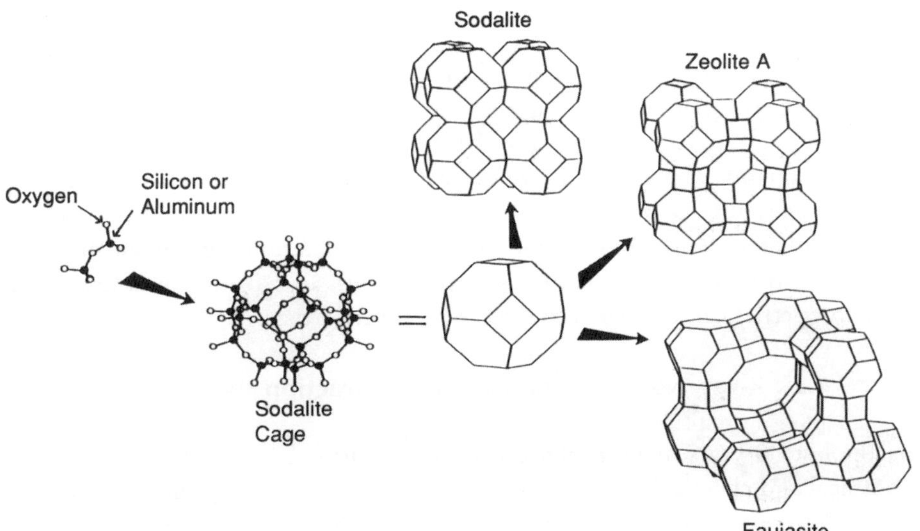

Figure 7.2 Schematics of zeolite frameworks. The synthetic faujasites are zeolites NaX and NaY. The determining factor between NaX and NaY is the Si/Al: NaX = 1.1 and NaY = 2.4. (From M. E. Davis, *Ind. Eng. Chem. Res.*, 30, 1675, 1991. Reprinted by permission of the American Chemical Society.)

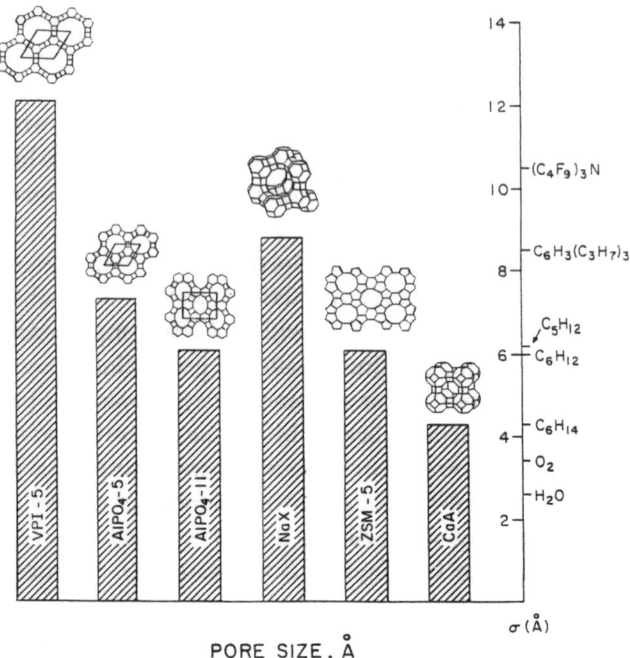

Figure 7.3   Correlation between the pore size of molecular sieves and the kinetic diameters of molecules. (From M. E. Davis, *Ind. Eng. Chem. Res.*, 30, 1675, 1991. Reprinted by permission of the American Chemical Society.)

Zeolites and aluminophosphate-based molecular sieves constitute a broad class of molecular sieves that have a wide variety of physicochemical properties, including variations in pore size (from about 4 to 13 Å), pore shape (circular or elliptical), dimensionality of pore system (one-, two-, or three-dimensional), presence or absence of cages, surface properties (hydrophilic or hydrophobic), void volume (up to 50%), and framework composition. Methods for determining framework composition and structure are briefly discussed below with an explanation of how these physical properties can be related to the chemical properties.

7.3   X-ray, Neutron, and Electron Diffraction

Over the last 30 years many analytical techniques have been used to characterize the physical and chemical properties of zeolitic materials. Among these, diffraction techniques have proved to be the most versatile and useful. The coherent scattering of radiation by zeolite crystals produces a "fingerprint" spectrum that contains information about crystal symmetry, crystallite shape and size, structural defects such as planar faults (which are common in these materials), and ultimately, the average atomic structure of the crystals. The diffraction of radiation by a single

          ZEOLITES AND MOLECULAR SIEVES   Chapter 7

crystal produces a three-dimensional data set in that the diffracted intensities are distributed at points in space that define a three-dimensional lattice. If a single crystal is replaced by a powder in the diffraction experiment, the three-dimensional distribution of intensities is effectively reduced to one dimension, resulting in a loss of information. Individual intensity distributions (commonly referred to as reflections) in the single crystal experiment are overlapped or superimposed in the powder experiment.

Most synthetic zeolites are crystalline powders with crystallite sizes in the 0.01 to 10 μm size range. These materials are too small for conventional single crystal X-ray and neutron studies, but are ideal for powder diffraction studies. Single crystal electron diffraction data from these small crystals are often helpful in indexing the powder diffraction data and identifying twinning and structural disorder. Recent advances in the synthesis of larger synthetic zeolite crystals and the availability of large single crystals of natural zeolites have made possible many single crystal X-ray and neutron studies. The advent of very high intensity synchrotron X-ray sources has opened the possibility of obtaining single crystal data from micron-sized single crystals, which were not amenable to study using conventional X-ray sources.

X-ray powder diffraction (XPD) studies have played the most important role in zeolite characterization because of the availability of powder diffraction instrumentation and the crystallite size considerations discussed above. XPD data also have significant commercial importance to the catalyst industry. X-ray *d*-spacings and their relative intensities are used to distinguish zeolitic materials in most zeolite composition of matter, synthesis, and modification patents. Since the first zeolite composition of matter patent issued in 1959, over 300 U.S. zeolite patents and many foreign zeolite patents have included XPD data.

Because of the degeneracy of XPD data, it is important to obtain the best possible data set with the available instrumentation. Careful attention must be paid to instrument alignment, calibration, and sample preparation, as described in many texts and papers on powder diffraction.[5–8] Bish and Post[8] present the most comprehensive review of modern XPD practices and provide an extensive list of references. Rohrbaugh and Wu[9] discuss the effects of crystallite size, morphology, sample preparation, instrument parameters, and other factors on zeolite XPD data. Most X-ray powder data are now collected in digital form using computer controlled diffractometers. The results of a computer-automated XPD experiment are a data set of incremental 2θ values, with an intensity measurement for each 2θ position. These are generally much preferred over analog powder patterns because computer processing of the digital data can extract more information with higher precision than manual measurement of analog patterns. Exceptions are focusing film cameras, especially Guinier-Hagg instruments, which are film cameras capable of excellent resolution. Resolution is extremely important in indexing powder diffraction data. The highest resolution XPD data available today is obtained from synchrotron sources. Examples of XPD work conducted at these facilities[10–13] is presented in facility yearbooks, such as the *National Synchrotron Light Source Annual Report.*

The information obtained about zeolites from the analysis of XPD data spans a range of uses from the preliminary identification of a zeolite—by visual comparison of patterns or tabulated data—to the accurate determination of atomic coordinates and their thermal parameters from a complete structural refinement. A brief discussion of several of these analyses and references to more detailed information is given below.

Identification of Zeolites

Many synthetic molecular sieves are white powders with aluminosilicate or aluminophosphate compositions. Their identification is often the first step in working with these materials. Many of the important properties of zeolites are dependent on their pore system, which is defined by the structure of their tetrahedral framework. Identification of the framework structure is the first step in the X-ray characterization of a zeolite. Much useful information on zeolite structure is provided by the *Atlas of Zeolite Structure Types*,[2] which is published and updated every few years under the auspices of the International Zeolite Association.

The zeolite structure type is usually identified by comparing the XPD data of the unknown material with reference X-ray data, which are available in journal articles, patent literature, technical books, and X-ray databases. Sources of reference X-ray data are tabulated in Table 7.1 with comments on their advantages and disadvantages. Zeolite samples often contain impurities, such as other zeolites or dense crystalline phases, which complicate comparisons of XPD data and make it important to exercise judgment in selecting reference data. The only way to assure proper and complete X-ray identification is to use reference data in which each

Source	Comments
Collection of Simulated XRD Powder Patterns for Zeolites[14]	Calculated (indexed) patterns from structure refinements. Patterns contain no peaks from impurities.
JCPDS powder diffraction files[15]	Some indexed experimental and calculated data. Contains most natural zeolites, but very few synthetic molecular sieves.
Journal literature	Data often not indexed. May contain impurity phases.
Patent literature	Contains much data not available in journal literature. Almost never indexed. May contain impurity phases.

Table 7.1   Sources of reference X-ray data.

   ZEOLITES AND MOLECULAR SIEVES   Chapter 7

X-ray reflection has been indexed (i.e., assigned a Miller Index) for a particular phase. Reference data sources such as the "Collection of Simulated XRD Powder Patterns for Zeolites"[14] and the *JCPDS Inorganic and Mineral Powder Diffraction Files*[15] are usually better than journal and patent sources which may not contain indexed XPD data.

Compositional and Phase Changes

Diffraction data is extremely sensitive to variations in the size, symmetry, and composition of the zeolite unit cell. These variations are usually related to changes in intensive variables such as temperature, pressure, composition, and stress. Changes in lattice parameters may be observed directly in the diffraction data as changes in the 2θ positions of reflections. Symmetry changes may be recognized by the splitting of reflections and the appearance or absence of certain classes of reflections, as defined by their Miller indices. Complete structure refinements of phases before and after a transformation may provide clues to the actual transformation mechanism. The following examples illustrate the use of diffraction data to characterize composition and phase changes.

Synthetic X and Y (faujasite) zeolites are used almost exclusively as commercial cracking catalysts. Catalytic properties of these materials are related to the Al framework content, which affects the magnitude of the cubic unit cell parameter a_0. Breck and Flanigen[16] observed systematic changes in a_0 in Ca-exchanged X and Y zeolites as a function of the unit cell Al content. They suggest that lattice constant discontinuities near 62 and 77 Al atoms per unit cell (Figure 7.4, left) are possibly related to the movement of Ca ions to different pore sites, resulting in a distortion of the zeolite framework. A study of synthetic sodium faujasites by Dempsey et al.[17] revealed a similar correlation with discontinuities at 65 and 80 Al atoms per unit cell (Figure 7.4, right). The authors suggest that the discontinuities may be related to phase changes that result from changes in the silicon–aluminum ordering scheme.

A study[18] of the displacive phase transformation in ZSM-5 using powder diffraction revealed a reversible, displacive, orthorhombic to monoclinic transformation in samples with SiO_2/Al_2O_3 ratios that range from about 70 to greater than 3000. As shown in the XPD data (Figure 7.5), the orthorhombic, as-synthesized material transformed to the monoclinic hydrogen form after calcination and then back to the orthorhombic form after NH_3 sorption. Single crystal X-ray studies[19, 20] reveal that the transformation is realized through a complex displacement of framework atoms substantially enlarging the sinusoidal channels in the framework for the diffusion of p-xylene. The orthorhombic to monoclinic transformation in ZSM-5 produces two sets of monoclinic twin domains, which appear as planar microstructures in the optical images. Application of a uniaxial mechanical stress to the crystal causes one of the two twin domains to grow at the expense of the other. A reversal in the direction of the shear stress causes the domains to exhibit an opposite growth pattern. This type of behavior is called elastic hysteresis, and crystals exhibiting this behavior are called ferroelastic. The transformation from the untwinned orthorhombic ZSM-5 to the ferroelastic monoclinic ZSM-5 is an example of a ferroelastic phase transformation. Because the ferroelastic transformation results in a distinct physical change in the structure of the crystal, both sorption and catalytic properties may be affected. Detailed X-ray structure refinements of the orthorhombic and monoclinic ZSM-5 frameworks[19, 20] revealed changes in the dimensions of the sinusoidal channel from 5.3 by 5.6 Å in the orthorhombic form to 5.8 by 5.9 Å in the monoclinic form.

Reconstructive phase transformations are also amenable to study using diffraction techniques. The thermal history[21] of synthetic Na TMA-E (Figure 7.6) reveals a transformation of a sodalite-type phase at about 360 °C. Changes in the 2θ positions of several X-ray reflections before the transformation indicate that substantial changes in the lattice parameters precede the transformation. The sharp transformation observed in the X-ray data indicates no loss of crystallinity and suggests a topotactic transformation, in which the solid product is formed in one or several crystallographically equivalent orientations relative to the parent crystal, with the transformation able to proceed throughout the parent crystal. The solution of the TMA-E framework structure suggested a transformation to the sodalite structure that involves an acid–base reaction proceeding in characteristic loops through the structure.

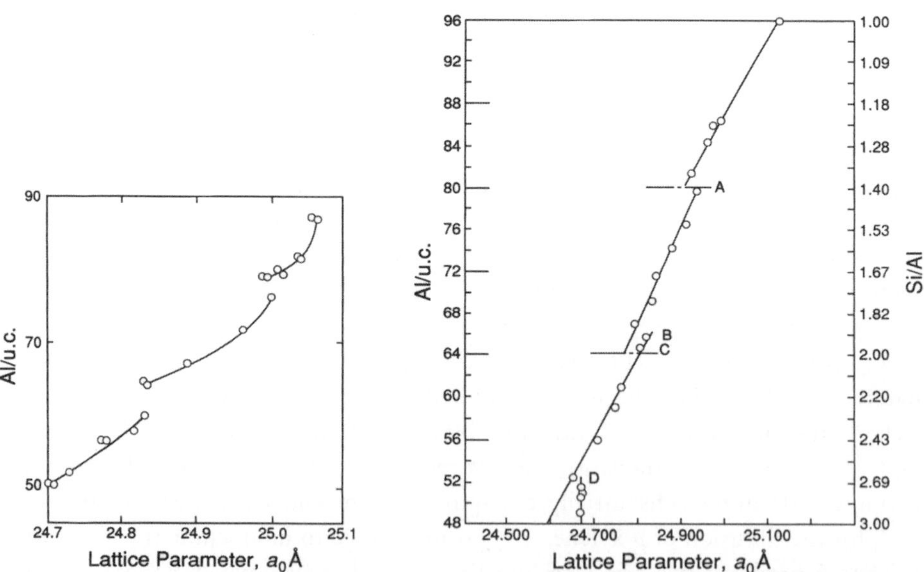

Figure 7.4 (left) The relationship between the cell constant, a_0, and the Al per unit cell, Al/u.c., of dehydrated forms of calcium-exchanged X and Y zeolites. (right) The number of Al per unit cell and the Si/Al ratio versus a_0 for hydrated synthetic sodium faujasites. (From Reference 17. Reprinted by permission of the American Chemical Society.)

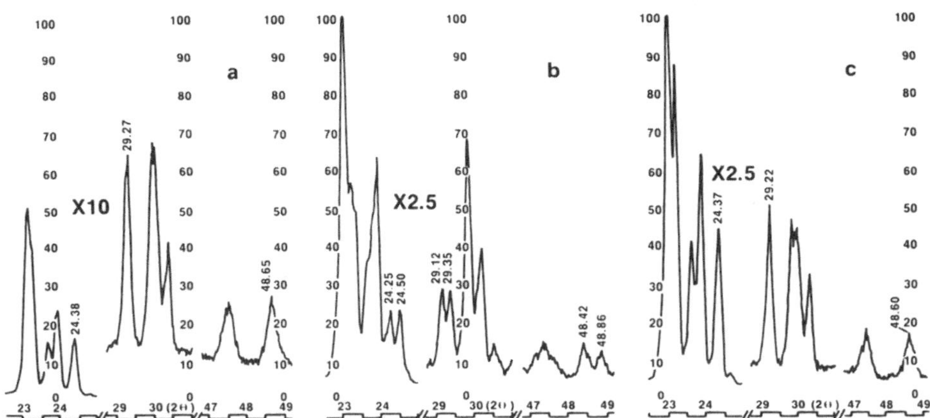

Figure 7.5   X-ray diffraction patterns of ZSM-5: (a) as-synthesized, (b) hydrogen form, (c) after loading with NH₃. (From Reference 18. Reprinted by permission of the American Chemical Society.)

Structure Determination by Diffraction Techniques

A knowledge of zeolite crystal structure and, in particular, pore architecture is highly desirable, because most catalytic applications for zeolites depend at least in part on their shape selectivity. X-ray diffraction has been the primary tool for investigating zeolite crystal structures, as evidenced by the several hundred structure

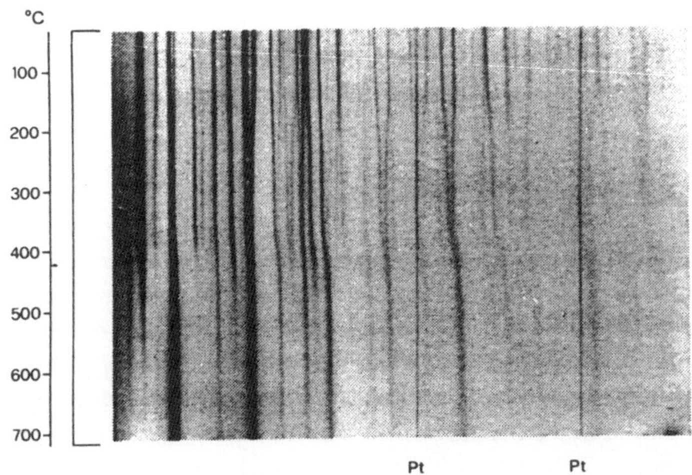

Figure 7.6   Guinier-Lenne X-ray film pattern of zeolite (Na, TMA)-E heated in air at a rate of 0.5 °C per minute. Significant changes in the 2θ positions (horizontal axis) of reflections are evident at about 450 °C. Reflections from an internal Pt standard are labeled. (From Reference 21. Reprinted by permission of the authors and Academic Press, Inc.)

determinations and refinements that appear in the published literature. Neutron diffraction studies, though less common, have played an important role in studying hydrogen bonding associated with water or hydrocarbons occluded in the pore systems. In the simplest case, the pore content of the zeolite is removed by various means, such as calcination, leaving only the structure of the tetrahedral framework. Determining the location of cations, organics, and water occluded in pore systems is often more difficult. Occluded species may not order in the pores or they may possess point and translational symmetries which are incompatible with those of the framework. These types of problems are especially common in synthetic materials.

Depending on the size of the zeolite crystals, two different approaches have generally been adopted for determining the structure of their tetrahedral framework. Zeolite crystals larger than about 50 μm on edge may be studied using single crystal techniques. It is often possible to collect several thousand unique X-ray intensities from one of these single crystals. The phase associated with each reflection is usually obtained by direct methods using one of several available computer programs. As has been pointed out in the recent X-ray structure determination textbook by Stout and Jensen[22]: "The methods have been automated to the extent that they are commonly used as 'black box' techniques in which the raw data go in at one end and the essentially solved structure appears at the other."

However, smaller crystallite "powdered" samples, which include most synthetic zeolite materials, require X-ray powder diffraction techniques that resolve an order of magnitude fewer intensities, which is often insufficient for the direct solution methods discussed above. This situation has provided a fertile field of research for zeolite crystallographers and has led to the development of several nontraditional methods for solving zeolite framework structures. Such structure determination techniques are beyond the scope of this chapter (McCusker[23] provides a recent review of this field).

7.4   High-Resolution Electron Microscopy

Information from high-resolution electron microscopy (HREM) bridges the gap between short range structure data obtained through NMR and long range crystallographic order data obtained through X-ray and neutron diffraction. Thus, HREM is the ideal tool for examining microtextures that result from growth faults and twinning, both common in synthetic zeolites. A recent HREM study[24] of ZSM-20 revealed a faulted block intergrowth of cubic and hexagonal stacking of faujasite 111 layers (Figure 7.7). The cubic and hexagonal sequences were not random, but clustered, with extended slabs of the two frameworks intermixed. The conclusion was that this planar faulting does not block the pore system, but does modify the pore tortuosities and geometries. Similar HREM studies exist[25, 26] of ZSM-5/ZSM-11 intergrowths and disordered zeolite beta.

7.5 Solid State NMR Spectroscopy

NMR spectroscopy has revolutionized the study of atomic ordering in molecular sieves. Through the application of multinuclear NMR spectroscopy, the coordination and element composition of the second coordination (the first coordination is oxygen) for various elements can be ascertained, as described below. See Fyfe[27] for a further discussion of the theory and implementation of solid state NMR and Englehard[28] for the application of solid state NMR to the study of molecular sieves.

Framework Composition

The bulk composition of a solid sample can be determined by any one of several different techniques, such as X-ray fluorescence. More commonly, the zeolite or molecular sieve is dissolved, and the soluble components are analyzed using atomic absorption (AA) or inductively coupled plasma (ICP) spectroscopy. Corbin et al.[29] provide a good comparison of bulk chemical analysis techniques.

Since NMR can yield information concerning the coordination and composition of second-nearest neighbors, it can be used to ascertain framework composition. The framework composition need not be the same as the bulk composition; non-framework species in the bulk can be identified and by NMR, then eliminated from the framework composition calculation (see below). Li et al.[30] synthesized a series of zeolites (hexagonal faujasites) that have the same framework topology, with a varying silicon/aluminum ratio. The ^{29}Si NMR spectra from these zeolites (see

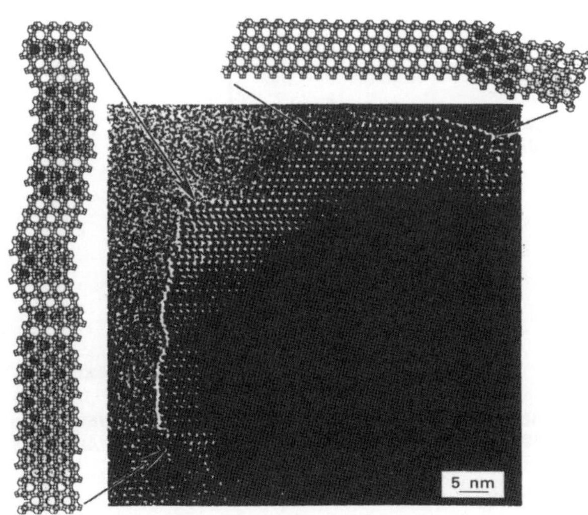

Figure 7.7   **Electron micrograph of a typical ZSM-20 crystallite aggregate. (Reprinted by permission of The Royal Society of Chemistry. Special thanks to M. M. J. Treacy)**

Figure 7.8) show peaks at –84, –90, –95, –101, and –106 ppm, which correspond to the silicon coordinations of Si(4Al), Si(3Al, 1Si), Si(2Al,2Si), Si(1Al,3Si) and Si(4Si), respectively (the oxygens are deleted for notational simplicity). Also, amorphous SiO_2 produces a broad ^{29}Si NMR peak at –109 ppm. These spectra are deconvoluted to obtain the relative area ratios of the six environments, and the framework composition of each sample can be calculated by applying the following formula:

$$\left(\frac{Si}{Al}\right)_f = \frac{\sum\limits_{n=0}^{4} A_{Si(nAl)}}{\sum\limits_{n=0}^{4} 0.25\, n A_{Si(nAl)}}$$

where $A_{Si(nAl)}$ is the relative area of the ^{29}Si NMR peak assigned to the environment Si(nAl). The peak at –109 ppm does not contribute to this calculation since it is from non-framework silicon. As calculated from the ^{29}Si NMR data for hex(0), the bulk $(Si/Al)_b$ is the same as the framework $(Si/Al)_f$. In this case, there is no peak at –109 ppm and non-framework aluminum does not exist. In samples where $(Si/Al)_f > (Si/Al)_b$, there must clearly be some non-framework aluminum. Figure 7.9 shows the ^{27}Al NMR spectra from a sample where $(Si/Al)_b = (Si/Al)_f$ and

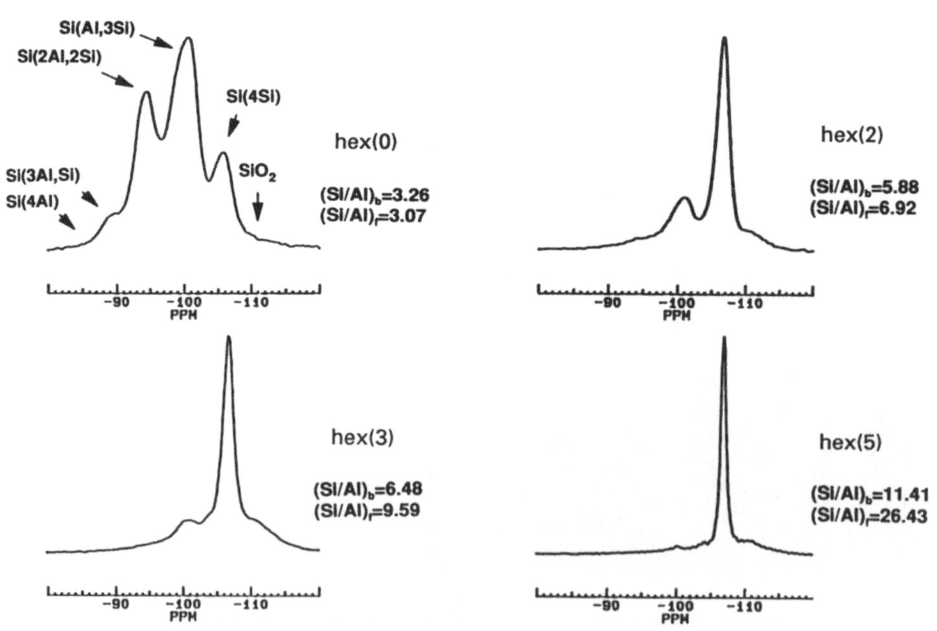

Figure 7.8 ^{29}Si MAS NMR spectra. $(Si/Al)_f$ is taken from NMR spectrum, and $(Si/Al)_b$ from bulk chemical analysis.

  ZEOLITES AND MOLECULAR SIEVES Chapter 7

$(Si/Al)_b < (Si/Al)_f$. The peak at about 0 ppm is from octahedral (i.e., non-framework) aluminum, proving that the existence of non-framework aluminum is observable using ^{27}Al NMR.

Tetrahedral Atom Ordering

Since the chemical shift of an NMR-active element is sensitive to the type of atom in its second coordination, it is possible to propose the ordering of the tetrahedral atoms in the available framework positions. For example, the chemical shifts from Al(4Si) and Al(4P) are in the neighborhood of 60 ppm and 40 ppm, respectively. Saldarriaga et al.[31] report that the chemical shifts for silicon, aluminum, and phosphorus from the silicoaluminophosphate molecular sieve SAPO-37 are –90, 38, and –25 ppm, respectively. (In each spectrum there was a single NMR peak indicating that only a single environment existed for each element.) These chemical shifts are indicative of Si(4Al), P(4Al), and Al(4P) or Al(3P,1Si). Since these elements are homogeneously distributed over the framework sites (with one peak in each spectrum), the aluminum peak must be from Al(3P,1Si) in order to provide the connectivity between silicon and phosphorus. These data, in combination with the measured bulk chemical composition (4Al : 3P : 1Si), give the tetrahedral atom ordering shown in Figure 7.10.

New Developments

Pines et al.[32] have recently developed double rotation (DOR) NMR. Magic angle spinning NMR (MAS-NMR) is effective for studying spin ½ nuclei, such as ^{29}Si.

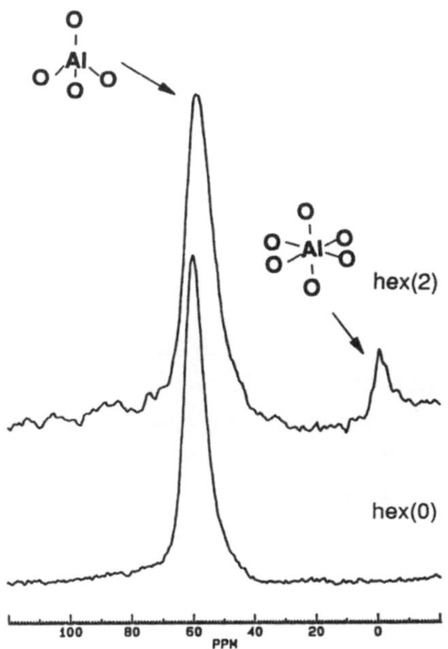

Figure 7.9   **^{27}Al MAS NMR spectra.**

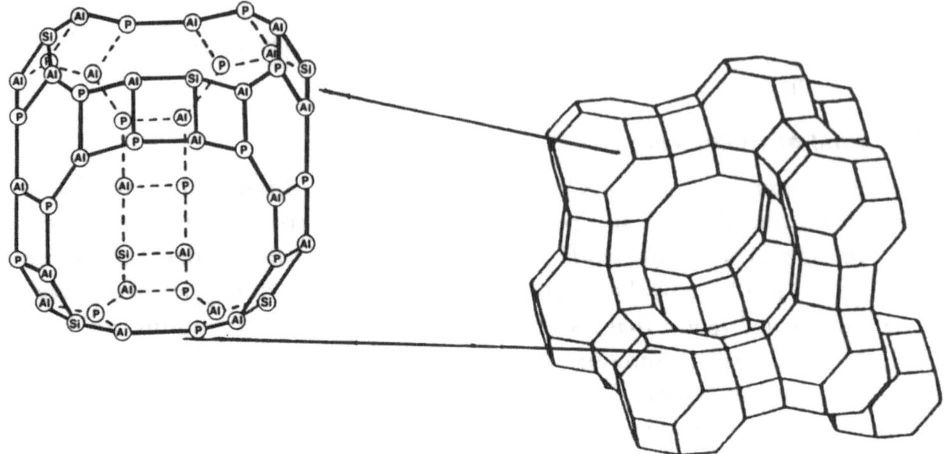

Figure 7.10 Ordering of Si, Al, and P in SAPO-37. The additional oxygen bond for each T-atom on the labeled portion of the diagram has been removed for notational simplicity.

However, for quadrupolar nuclei, such as ^{27}Al, spectral broadening from anisotropic second-order quadrupolar interactions cannot be eliminated by MAS-NMR. To achieve line-widths comparable to spin ½ nuclei, DOR NMR must be used. Using this technique, in which the sample is rotated simultaneously about two axes, Pines and co-workers[33, 34] have obtained high-resolution NMR spectra of ^{17}O and ^{27}Al.

As shown above, NMR spectroscopy complements diffraction techniques in providing information on molecular sieve structures. Fyfe et al.[35] have recently shown that ^{29}Si two-dimensional NMR can be observed using natural-abundance ^{29}Si in high-silica molecular sieves. Thus, information concerning the three-dimensional framework connectivities of silicon sites can be established through NMR studies.

7.6 Adsorption

Void Volume

The total void volume in molecular sieves is commonly measured by collecting water adsorption isotherms. Small molecules like water, H_2, and He can access all void spaces in these materials. Since most molecular sieves contain some type of occluded material, such as water, the samples must be activated prior to adsorption experiments. This typically involves heating the sample in a vacuum until the remaining solid consists of nothing but inorganic oxides. The sample is then activated, and the adsorption of water, or the volumetric displacement of He or water (pycnometry), yields the void volume of the sample. The void fractions of zeolites and molecular sieves are approximately 50% or less.

Kinetic Diameter, Å	Adsorbate	Capacity, cm^3/ga		
		AlPO$_4$–11	AlPO$_4$–5	VPI-5
3.46	O$_2$	0.085	0.146	0.228
4.30	*n*-hexane	–	0.139	0.198
6.00	cyclohexane	0.070	0.145	0.156
6.20	neopentane	0.028	0.137	0.148
8.50	triisopropylbenzene	–	0.021	0.117

Table 7.2 Adsorption capacity[36] of molecular sieves at P/P$_0$ = 0.4.

Pore Size

As previously mentioned, molecular sieves are unique in that they contain pores in the size range of small molecules. These pore sizes can be measured using adsorption methods, as described below.

Table 7.2 lists the adsorption capacity of AlPO$_4$-11, AlPO$_4$-5 and VPI-5 (see Figure 7.3 for structures). These molecular sieves contain unidimensional pores that are 6.3 by 3.9 Å elliptical, 7-3 Å circular, and 12.1 Å circular, respectively. The molecular sieve effect is illustrated by AlPO$_4$-11, which shows a similar adsorption capacity for oxygen and cyclohexane because their molecules are both smaller than the pore size of AlPO$_4$-11. Since a neopentane molecule is larger than the pore size of AlPO$_4$-11, it is not adsorbed in any significant quantity. Similarly, AlPO$_4$-5 adsorbs oxygen and all hydrocarbons listed in the table except triisopropylbenzene, since it is larger than the pore size of AlPO$_4$-5. The capacity of AlPO$_4$-5 for the other four adsorbates is roughly the same. The adsorption capacity of VPI-5 declines with increasing molecular size. This behavior, unique to VPI-5, is caused by the packing of small molecules in the 12.1 Å pores of VPI-5. This does not occur in AlPO$_4$-11 and AlPO$_4$-5 because the probe molecules listed are of sufficient size that only one molecule can be placed across the pore diameter. However, VPI-5 pores are large enough to accommodate two or more small molecules such as oxygen or hexane simultaneously. The larger molecules such as neopentane can be accommodated only singly. Molecules sufficiently large not to be adsorbed in VPI-5 are not listed in Table 7.2 because the vapor pressure of such molecules low enough to render them useless for vapor phase adsorption experiments.

The use of probe molecules of varying size to determine the pore sizes of molecular sieves has been quite successful, particularly for pore sizes of less than 10 Å. However, there is only a limited selection of probe molecules for pores larger than 10 Å. High-resolution argon adsorption isotherms provide an alternative method for determining pore sizes in molecular sieves. Hathaway and Davis[37] have shown that the inflection point in the argon adsorption isotherm can be correlated with the pore size in molecular sieves. The use of argon is important, since it has no quadrupole moment. The advantages of this technique are speed (the use of a number of probe molecules is laborious) and the ability to probe a much broader range in pore size than is possible with volatile probe molecules. The disadvantage is the lack of a precise mathematical relationship that can convert the isotherm to a numerical value of pore size. However, since many molecular sieves of known pore size are available, the isotherm position (in terms of P/P_0) can be calibrated to known pore sizes. Figure 7.11 illustrates the argon isotherm from $AlPO_4$-11, $AlPO_4$-5, and VPI-5. These three materials practically span the pore size range in zeolites and molecular sieves (pore sizes smaller than $AlPO_4$-11 do exist).

7.7   Structure and Catalytic Behavior

Zeolites and other molecular sieves are useful as catalysts, and because of their structural uniformity, they provide opportunities to study structure–property relationships in a meaningful manner. Two aspects of the structure–catalytic property relationships are illustrated below.

Many catalytic applications[38] have utilized the ability of zeolites to discriminate molecules on the basis of their size. Following is a particularly good example of this feature: zeolites A and X (LTA and FAU, respectively) are first ion exchanged with

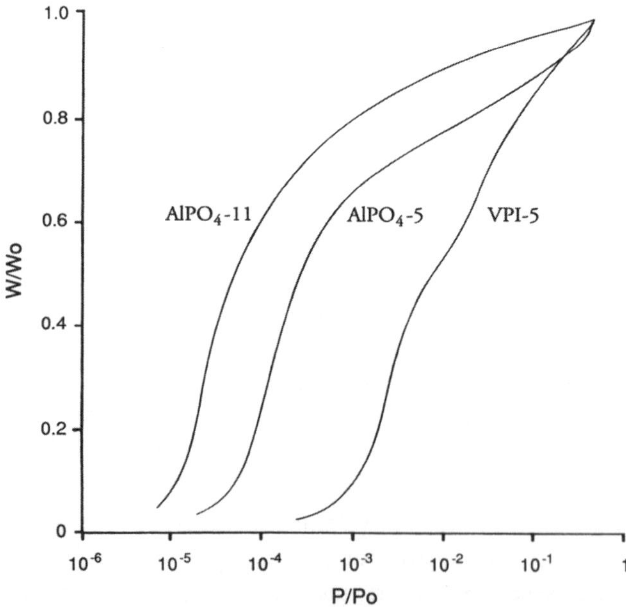

Figure 7.11   **High-resolution argon adsorption isotherms from molecular sieves at 87 K. (From Reference 36. Reprinted by permission of the American Chemical Society.)**

calcium salts [creating H^+ sites through the hydrolysis of the water of hydration that surrounds the calcium ions located within the zeolite, e.g., $Ca(H_2O)_x^{2+} \rightarrow Ca(H_2O)_{x-1}(OH)^+ + H^+$] and then contacted with primary and secondary butanol (*n*- and *i*-, respectively) in the vapor phase. Figure 7.12 illustrates the results obtained.[39] Both alcohols dehydrate to alkenes on CaX since these molecules are able to sorb into the 8 Å pores of the zeolite. However, only the primary alcohol is sufficiently small to be sorbed by CaA, due to its pore size of only 4–5 Å. Thus, it is clear that zeolites can produce reaction selectivities with sub-Ångström level control (in that the difference in molecule size between the primary and secondary alcohol is less than 1 Å) by imposing structural constraints on catalytic reaction pathways.

In addition to the structure–property relationships caused by spatial arrangements, number density relationships can be controlled. For example, each aluminum atom in the framework of ZSM-5 (see Figure 7.3) creates a negative charge that can be balanced by H^+ to give a solid acid catalyst. Haag[40] shows that the rate of *n*-hexane cracking is linearly related to the number of Al atoms in the ZSM-5 framework as well as to the intensity of the ^{27}Al-NMR signal. Thus, the number of H^+ sites (which are determined by the number of Al sites) correlates with the number of *n*-hexane molecules cracked.

7.8 Summary

Molecular sieves are crystalline inorganic oxides with well-defined structures that contain molecule-sized void spaces. Through the use of various techniques, the

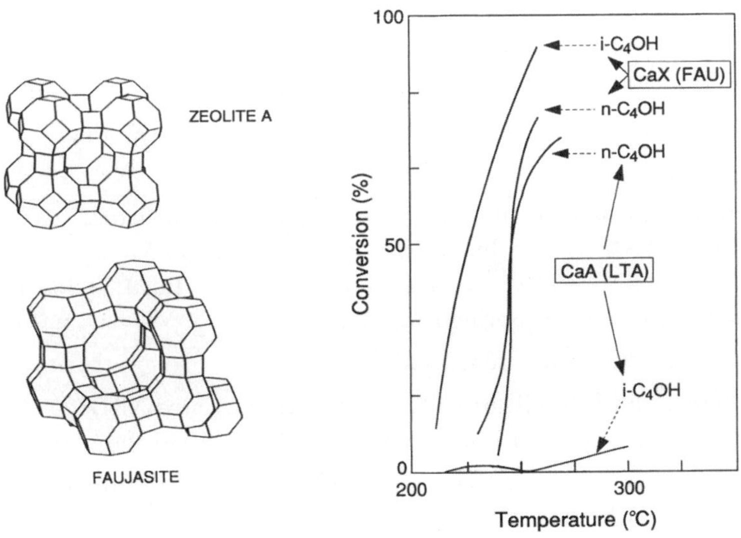

Figure 7.12 Dehydration of alcohols by zeolites. (After Weisz et al.[39])

physicochemical properties of these materials can be defined, and the structure–catalytic behavior relationships explored. These materials will continue to find widespread application because of their ability to organize and react molecules with Ångström-level specificity.

References

1   J. Newsam. *Mat. Sci. Forum.* **27/28**, 385, 1988.

2   W. M. Meier and D. H. Olson. *Atlas of Zeolite Structure Types.* Butterworth, London, 1987.

3   R. M. Szostak. *Molecular Sieves: Principles of Synthesis and Identification.* Van Nostrand-Reinhold, New York, 1989.

4   E. M. Flanigen, R. L. Patton, and S. T. Wilson. *Stud. Sur. Sci. Catal.* **37**, 13, 1988.

5   H. S. Peiser, H. P. Rooksby, and A. J. C. Wilson. *X-ray Diffraction By Polycrystalline Materials.* Chapman & Hall Limited, London and Reinhold, New York, 1960.

6   H. P. Klug and L. E. Alexander. *X-ray Diffraction Procedures for Polycrystalline and Amorphous Materials.* 2nd ed., Wiley, New York, 1974.

7   L. V. Azaroff and M. L. Buerger. *The Powder Method in X-ray Crystallography.* McGraw-Hill, New York, 1958.

8   D. L. Bish and J. E. Post. *Modern Powder Diffraction, Reviews in Mineralogy, Vol. 20.* The Mineralogical Society of America, Washington, D.C., 1989.

9   W. J. Rohrbaugh and E. L. Wu, *ACS Sym. Ser.*, **411**, 279, 1989.

10 S. Hulbert, N. Lazarz, and G. Williams. In *National Synchrotron Light Source Annual Report 1988.* BNL-52167 UC-400. (S. Hulbert et al., Eds.) National Technical Information Service, Springfield, VA, 1988.

11 Schlenker, J. L., Higgins, J. B., and E. W. Valyocsik. *Zeolites.* **10**, 293, 1990.

12 J. M. Bennett, R. M. Kirchner, and D. E. Cox. "The Crystal Structure of an As-Synthesized $AlPO_4$-16," In *National Synchrotron Light Source Annual Report 1987.* BNL-52131 UC-400 (S. Hulbert et al., Eds.) National Technical Information Service, Springfield, VA, 1987.

13 L. McCusker. *J. Appl. Cryst.* **21**, 305, 1988.

14 R. B. von Ballmoos and J. B. Higgins. "Collection of Simulated XRD Powder Patterns for Zeolites," In *Zeolites.* **10**, 313S–514S, 1990.

15 *Powder Diffraction File, Inorganic Phases.* JCPDS International Centre for Diffraction Data, Swarthmore, PA, 1987.

16 D. W. Breck and E. M. Flanigen. *Molecular Sieves.* Society of Chemical Industry, London, 1968, p. 47.

17 E. Dempsey, G. H. Kuhl, and D. H. Olson. *J. Phys. Chem.* **73**, 387, 1969.

18 E. L. Wu, S. L. Lawton, D. H. Olson, A. C. Rohrman, and G. T. Kokotailo. *J. Phys. Chem.* **83**, 2777, 1979.

19 H. van Koningsveld, J. C. Jansen, and H. van Bekkum. *Zeolites.* **10**, 235, 1990.

20 H. van Koningsveld. *Acta Cryst.* **B46**, 731, 1990.

21 W. M. Meier and M. Groner. *J. Solid State Chem.* **37**, 204, 1981.

22 G. H. Stout and L. H. Jensen. *X-ray Structure Determination, a Practical Guide.* 2nd ed., Wiley, New York, 1989, p. 248.

23 L. B. McCusker. *Acta Cryst.* **A47**, 297, 1991.

24 J. M. Newsam, M. M. J. Treacy, D. E. W. Vaughan, K. G. Strohmaier, and W. J. Mortier. *J. Chem. Soc., Chem. Commun.* 493, 1989.

25 J. M. Thomas and G. R. Millward. *J. Chem. Soc., Chem. Commun.* 1380, 1982.

26 J. M. Newsam, M. M. J. Treacy, W. T. Koetsier, and C. B. de Gruyter. *Proc. R. Soc. Lond. A.* **420**, 375, 1988.

27 C. A. Fyfe. *Solid State NMR for Chemists.* CFC, Guelph, Ontario, 1983.

28 G. Engelhard. *High-Resolution Solid-State NMR of Silicates and Zeolites.* Wiley, Chichester, 1987.

29 D. R. Corbin, B. F. Burgess, A. J. Vega, and R. D. Farlee. *Anal. Chem.* **59**, 2722, 1987.

30 H. X. Li, C. Y. Chen, M. J. Annen, J. P. Arhancet, and M. E. Davis. *J. Materials Chem.* **1**, 79, 1991.

31 L. S. Saldarriaga, C. Saldarriaga, and M. E. Davis. *J. Am. Chem. Soc.* **109**, 2686, 1987.

32 A. Samoson, E. Lippmaa, and A. Pines. *Mol. Phys.* **65**, 1013, 1988.

33 B. F. Chmelka, K. T. Mueller, A. Pines, J. Stebbins, Y. Wu, and J. W. Zwanziger. *Nature.* **339**, 42, 1989.

34 Y. Wu, B. F. Chmelka, A. Pines, M. E. Davis, P. J. Grobet, and P. A. Jacobs. *Nature.* **346**, 550, 1990.

35 C. A. Fyfe, H. Gies, Y. Feng, and G. T. Kokotailo. *Nature.* **341**, 223, 1989.

36 M. E. Davis, C. Montes, P. E. Hathaway, J. P. Arhancet, D. Hasha, and J. M. Garces. *J. Am. Chem. Soc.* **111**, 3919, 1989.

37 P. E. Hathaway and M. E. Davis. *Catal. Lett.* **5**, 333, 1990.

38 N. Y. Chen, W. E. Garwood, and F. G. Dwyer. *Shape Selective Catalysis in Industrial Applications.* Marcel Dekker, New York, 1989.

39 P. B. Weisz, V. J. Frilette, R. W. Maatman, and E. B. Mower. *J. Catal.* **1**, 307, 1962.

40 W. O. Haag. In *Proc. Sixth Int. Zeolite Conf.* (D. Olson and A. Bisio, Eds.) Butterworth, London, 1984, p. 466.

Alumina Pillared Clays: Methods of Preparation and Characterization

JEAN-RÉMI BUTRUILLE AND THOMAS J. PINNAVAIA

Contents

8.1 Introduction

Pillared clays are a relatively new class of microporous solid acids.[1] They are extensively studied as catalysts for many types of reactions, especially acid-catalyzed reactions such as cracking, alkylation, or isomerization.[2]

Pillared lamellar solids are intercalation compounds that meet two important criteria: (1) the gallery species are sufficiently robust to prevent gallery collapse upon dehydration and (2) the pillars are laterally spaced to allow for interpillar access by molecules at least as large as nitrogen. (See Pinnavaia and Kim[3] for the specific meaning of the term "pillared.")

Smectite clays are lamellar compounds with negatively charged layers. In order to achieve electroneutrality, the layer charge is compensated by interlayer cations such as Na^+ or Ca^{2+}. When in polar solvents, the layers swell and the interlayer cations become accessible and exchangeable. In 1955, Barrer and McLeod[4] discovered that the exchange of the smectite counter ions by means of highly charged, large cations yielded materials with a substantial internal microporous volume. The large cations function as pillars and prevent the collapse of the structure during outgassing (see Figure 8.1). This early work made use of organic cations, such as tetraalkylammonium ions, that had the disadvantage of being unstable above

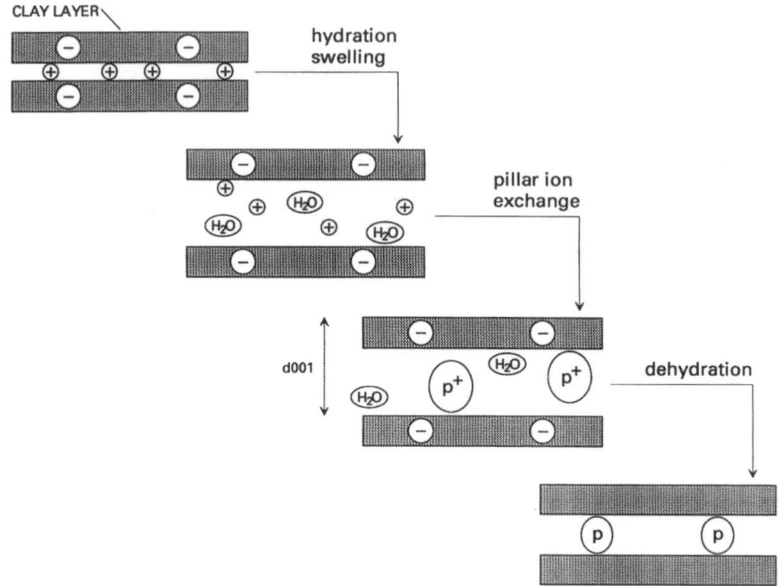

Figure 8.1   General scheme for the synthesis of pillared clays. Depending on the nature of the pillaring reagent, the final pillaring species may be charged, as in intercalate derived from NR_4^+ cations, or neutral, as in metal oxide pillared clays derived from inorganic polycations.

300 °C. The introduction of inorganic pillars in the late 1970s afforded much more robust intercalated products,[5–7] some of which were stable above 500 °C. The most effective pillars are metal oxide aggregates derived from polycations of aluminum,[5–7] zirconium,[8] chromium,[9] iron,[10, 11] and titanium.[12] The nature of the polycation and the host clay gives rise to different catalytic properties.

The focus here is on the most commonly used polycation pillaring agent, the Keggin-like ion $[Al_{13}O_4(OH)_{24}(H_2O)_{12}]^{7+}$, which is generated by the base hydrolysis of an aluminum chloride solution. The pillared clays obtained using this cation exhibit Brönsted or Lewis acidity or both, depending on the nature of the clay. This acidity is believed to be a consequence of the dehydration/dehydroxylation reaction that occurs upon calcination of the intercalated polycation at elevated temperatures[2]:

$$[Al_{13}O_4(OH)_{24}(H_2O)_{12}]^{7+} \longrightarrow 6.5Al_2O_3 + 7H^+ + 20.5H_2O$$

The physical properties of pillared clays that most significantly affect their catalytic behavior are porosity and acidity. These properties, essential to catalytic performance, are best characterized using adsorption and desorption techniques. Other useful tools for the characterization of pillared clays include X-ray diffraction, elemental analysis, electron microscopy, and MAS-NMR spectroscopy, which are discussed at length in other chapters of this book.

8.2 Synthesis Methods

As is true for many other heterogeneous catalysts, the synthesis method can greatly affect the catalytic properties of alumina pillared clays. Differences in porosity or acidity depend on the method used to form the pillaring agent, the method used to dry the products, and other processing variables.

The first step in pillared clay synthesis is the preparation of the pillaring agent. Essentially two methods have been used to prepare the $[Al_{13}O_4(OH)_{24}(H_2O)_{12}]^{7+}$ cation. The first method[13] is the base hydrolysis of an Al^{3+} solution at a $OH:Al^{3+}$ molar ratio between approximately 2.0 and 2.4, with a value of 2.4 preferred. The second method is to use a commercially available aluminum chlorohydrate solution such as Chlorohydrol®, a product of the Reheis Chemical Company (Berkeley Heights, N.J.). Aluminum chlorohydrate solutions are formed by the dissolution of Al metal in a solution of $AlCl_3$. The dilution level and aging of the pillaring solution determine the properties of the pillared product.[14] The optimum dilution level for Chlorohydrol is 0.06 M. Lower dilutions levels do not allow sufficient depolymerization, whereas higher dilution levels cause excess depolymerization. To obtain optimum pillared clay properties, an aging time of 10 days is recommended for the diluted Chlorohydrol pillaring solution.

In order to purify a naturally occurring smectite clay (obtainable from the Source Clay Minerals Repository at the University of Missouri, Columbia, Mo.), the clay should be treated with 1.0 N NaCl to ensure complete conversion to the swellable Na^+ exchanged form. Major impurities (such as quartz, carbonates, and iron oxides) are removed by sedimentation. Typically, a 1–2 wt % suspension of the Na^+ form of the clay is poured into a graduated cylinder and allowed to stand overnight before the fraction of clay remaining in the upper three-quarters of the suspension is siphoned off. The clay can then be slurried in a sodium acetate/acetic acid buffer (pH = 5.0) to remove the remaining carbonate impurities; it can also be slurried in citrate/dithionite to remove traces of iron oxide. These latter treatments are not essential for obtaining a pillared clay because sedimentation is effective in removing most of the dense impurities.

A suspension of Na^+ clay in water and a solution of the pillaring agent are then mixed by adding the clay suspension one drop at a time to the pillaring agent solution or by some other suitable method. It is normal to use an excess of pillaring agent with regard to the cationic exchange capacity of the clay (approximately 15 mmol Al^{3+}/meq of clay). Excess chloride salt will be removed from the clay by repeatedly suspending the clay in de-ionized water and centrifuging the mixture. The washing process frees the clay of excess electrolyte, which results in the flocculation of the clay layers.

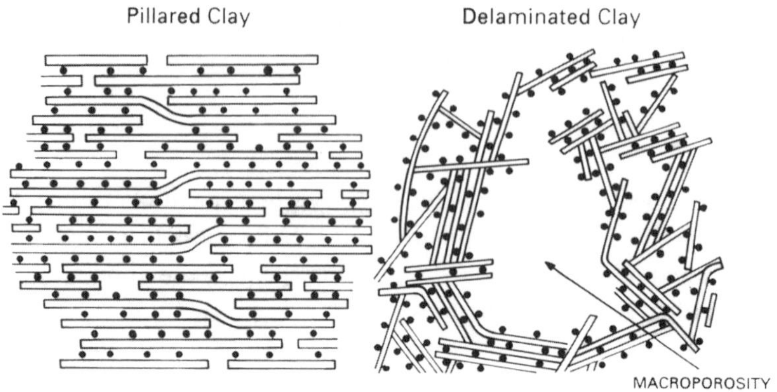

Figure 8.2   A layer aggregation model for pillared clay and delaminated clay.

The product can be either freeze-dried or air-dried. Air drying favors face–face orientation of the layers, whereas freeze drying tends to preserve the structure of the flocculated clay, which exhibits substantial edge–face orientations.[15] The morphology of the clay layers also plays a decisive role in the layer ordering. Layers having a small aspect ratio, such as Laponite (a synthetic clay produced by Laporte Industries, Ltd.), tend to organize by edge–face interactions, which results in a house-of-cards type of structure. On the other hand, typical smectites, such as montmorillonite, have large platelet diameters and organize in face–face structure (Figure 8.2). When no long range face–face order can be observed using X-ray diffraction, the material is said to be delaminated. Delaminated clays and pillared clays have very different pore structures.

Calcination of the air-dried Al_{13}–smectite reaction product at 350 °C converts the material to a stable metal oxide pillared derivative. If the product is not calcined soon after preparation, the pillared structure will eventually collapse through intracrystalline hydrolysis of the aluminum polycation. Depending on the humidity and temperature, this decomposition may occur over periods of weeks to months. Thus, calcination is an important step in producing an alumina pillared clay that remains stable over long storage periods at ambient conditions. The calcination process normally is carried out in an oven or in a tube furnace under a flow of air. Heating rates are typically in the range of 1 to 10 °C per minute; the heating time at 350 °C is at least 3 hours.

8.3   Properties of Pillared Clays

X-ray Diffraction Pattern

The quickest way to determine whether a pillar intercalation procedure was successful is to record an X-ray diffraction pattern of an oriented film of the product by allowing a suspension of the product to evaporate on a glass microscope slide.

ALUMINA PILLARED CLAYS...   Chapter 8

A basal spacing of about 18 Å gives immediate indication that pillaring occurred, since the layer thickness of a smectite is 9.5 Å and the size of the Keggin-like $[Al_{13}O_4(OH)_{24}(H_2O)_{12}]^{7+}$ ion is about 9.0 Å. Oriented film samples favor 001 Bragg reflections. The width of the X-ray diffraction peaks at one-half their maximum height can indicate the crystallinity of the pillared clay. Moore and Reynolds[16] discuss the different causes of line-broadening. Particle-size broadening occurs as a consequence of the small size of clay crystallites. The diffraction peak width can be used to give a quantitative estimate of the particle size—or more precisely, the size of the scattering domain—by the Scherrer equation:

$$L = \frac{\lambda K}{\beta \cos \theta}$$

where L is the mean crystallite dimension in Ångströms along the c-axis; K is a constant very near unity; and β is the width of a 2θ reflection at half of the peak height expressed in radians.

There are other causes of X-ray line-broadening. Smectite clays exhibit a turbostratic stacking, which means that the layers stack flat (face–face) on each other, but without alignment of the ab planes. This defect can contribute to variations in basal spacing and increase the line-broadening of the 001 reflection. Also, since hydrated pillars hold the layers apart, differences in pillar density or hydration can lead to a distribution of basal spacings. This distribution may correlate to the density of the pillars and thus to the cationic exchange capacity (CEC) of the clay. Charge localization on the layer can also be an important factor in determining pillar distribution. For example, the uncalcined pillared form of a fluorohectorite has a relatively large and localized charge density of 140 meq/100 g and exhibits much sharper XRD peaks than an analogous Wyoming montmorillonite, which has a low and delocalized layer charge of 75 meq/100 g (see Figure 8.3). Figure 8.3 also shows that the Al_{13} reaction products of clays with very small particle sizes (e.g., Laponite) can exhibit an amorphous X-ray diffraction pattern. As previously discussed, the absence of X-ray diffraction peaks shows that no long range face–face layer aggregation is present and that the material is an edge–face delaminated clay.[17]

Elemental Analysis

Elemental analysis is used to determine the amount of aluminum per unit cell that has been incorporated into the clay by pillaring. It is first necessary to obtain a satisfactory analysis of the starting clay.[18] Several techniques can be used to solubilize the sample for chemical analysis, such as fusion with lithium metaborate ($LiBO_2$) or dissolution in an $HF/HCl/HNO_3$ mixture with a microwave digester. The composition is then analyzed by inductively coupled plasma emission or atomic absorption. Newman and Brown[18] explain the procedure used to obtain a unit cell formula from the elemental analysis (it is beyond the scope of this book). Difficulties in determining the unit cell composition may be caused by the presence

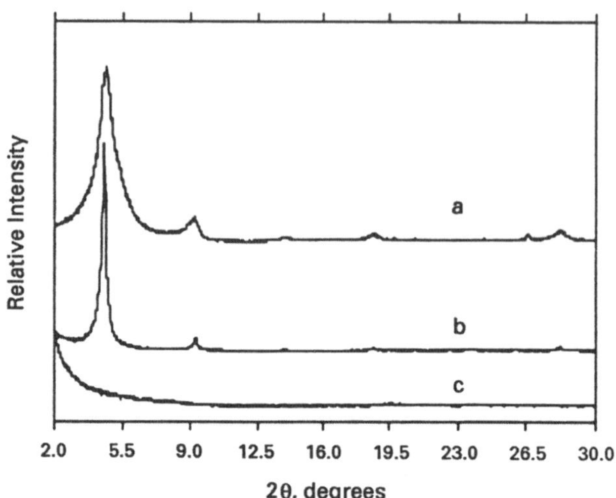

Figure 8.3 XRD of: (a) uncalcined Al_{13} pillared Wyoming montmorillonite; (b) uncalcined Al_{13} pillared fluorohectorite; and (c) delaminated Al_{13} Laponite.

of mineral impurities, such as quartz, kaolin, and other fine grain materials, and by the inaccurate subtraction of such impurities from the bulk analysis. If the clay is carefully purified before use, the impurity phases are usually low enough that their contribution can be disregarded. The complete unit cell formula is not necessary for determining the amount of aluminum incorporated for one $O_{20}(OH)_4$ unit. The quantity can also be obtained from the difference in the aluminum content before and after the pillaring reaction. It is given by the following simple formula:

$$N_{Al} = N_{Si}\left(\frac{Al_A}{Si_A} - \frac{Al_B}{Si_B}\right)$$

where N_{Al} and N_{Si} are the number of incorporated aluminum atoms and the number of layer silicon atoms, respectively, for one $O_{20}(OH)_4$ unit of the host clay; Al_A and Si_A are the relative molar amounts of aluminum and silicon, respectively, after the pillaring reaction; and Al_B and Si_B are the relative molar amounts of aluminum and silicon, respectively, before the pillaring reaction.

The amount of aluminum incorporated often is related to the charge density of the clay and thus with the CEC, as shown in Figure 8.4. This correlation requires that the same synthesis procedures are used to prepare each sample.

Electron Microscopy

The structure of pillared clays can be confirmed by lattice imaging using high-resolution transmission electron microscopy (HR-TEM), though this is not a routine technique. Basal spacings observed using HR-TEM agree with X-ray diffraction

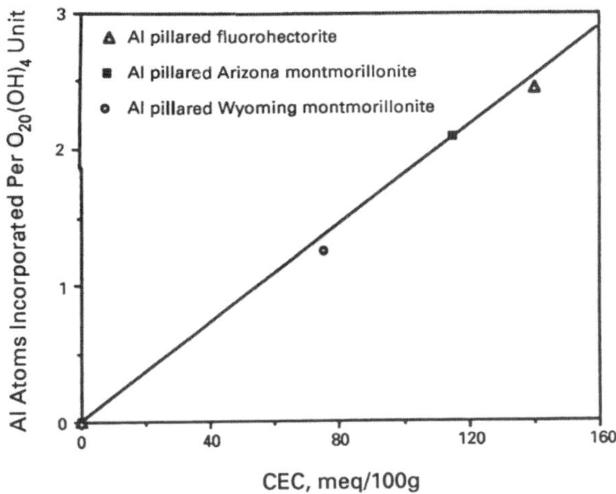

Figure 8.4   **Relationship between the amount of aluminum incorporated into the pillared clay and the CEC of the starting clay.**

(XRD) results.[19] HR-TEM images can provide information on the long-range ordering of the layers and on defects in these materials. HR-TEM is particularly useful for delaminated clays; XRD provides only limited information on structure. For example, HR-TEM was used by Occelli and Lynch[20] to determine that very small pillared regions that contain only a few face–face layers are present in delaminated Laponite. They also were able to observe poorly crystallized alumina.

Pore Structure by Adsorption-Desorption Techniques

Adsorption–desorption isotherms of probe molecules can provide quantitative information on the pore structure of pillared clays. N_2 adsorption is mainly used to obtain the specific surface area of the sample. A nonpillared clay is usually found to have a surface area of less than 50 m^2/g, whereas a pillared clay has a surface area in the 200–400 m^2/g range. The surface area is typically obtained from the adsorption branch by applying the BET equation. However, for microporous solids such as pillared clays, the BET equation does not apply over the usual partial pressure range of 0.05–0.25; instead, Langmuir's treatment is used by some researchers. The Langmuir equation is derived from a very simple model of monolayer adsorption that does not take into account the lateral interactions between adsorbate molecules; but the equation usually fits surprisingly well to the adsorption isotherm of microporous solids. However, this does not mean that the Langmuir model actually describes the adsorption process in the micropores. Since the Langmuir and BET models of multilayer adsorption are very different, there is no point in comparing the surface area values derived from these two models. It is customary for researchers to report BET surface areas for microporous solids such as pillared clays when reporting the BET surface areas of other solids; but these BET formulations

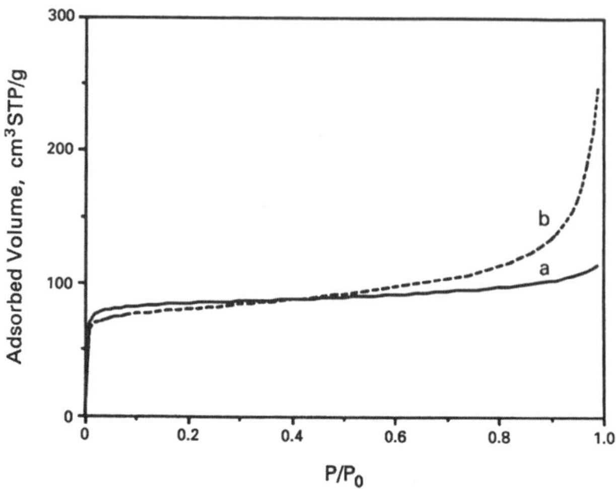

Figure 8.5 N$_2$ adsorption isotherms: (a) calcined (350 °C) alumina pillared Wyoming montmorillonite prepared from base hydrolyzed Al^{3+} (OH/Al = 2.4); and (b) calcined (350 °C) alumina pillared beidellite prepared from aluminum chlorohydrate.

are only accurate over the valid partial pressure range evidenced by a correlation coefficient near one. For pillared clays, the BET equation is usually valid when the value P/P$_0$ is between 0.01 and 0.10.

The presence of micropores is indicated by the validity of the Langmuir equation; however, t-plot or α_s-plot methods are more useful, since they not only indicate the presence of micropores, but also permit their quantification.[21] The t-plot and α_s-plot methods compare the adsorption isotherm of a porous solid with a standard adsorption isotherm for the same adsorbate on a nonporous solid. The t-plot shows a sample's adsorbed volume versus the statistical thickness of the adsorbed layer on the nonporous reference. The BET theory is used to help provide the values of the standard adsorbate layer statistical thickness, t. The BET theory is not used in the α_s-plot method. The α_s values are the ratios of adsorbed molecules per unit area at various P/P$_0$'S to adsorbed molecules per unit area at a reference P/P$_0$ of 0.4. For a non-microporous solid, the plot of adsorbed volume versus t or α_s is a straight line passing through the origin. The slope of the straight line is proportional to the surface area. For microporous solids such as pillared clays, the t-plot exhibits two slopes. Figures 8.5 and 8.6 display two N$_2$ adsorption-desorption isotherms and their treatment using the t-plot method. In the first region of the t-plot, the first few data points are fitted to a straight line passing through the origin, with the slope of this line yielding an equivalent surface area. If the adsorption data are reported in cm^3 at STP conditions, this surface area is given by

$$S = \sigma K b_{t_0}$$

                    ALUMINA PILLARED CLAYS...   Chapter 8

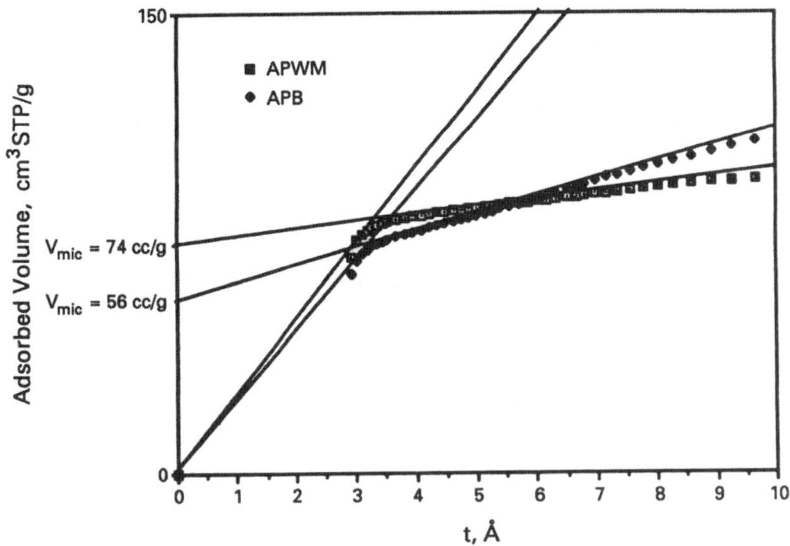

Figure 8.6　t-plots of N_2 adsorption on: calcined (350 °C) alumina pillared Wyoming montmorillonite (APWM) prepared from base hydrolyzed Al^{3+} (OH/Al = 2.4); and calcined (350 °C) alumina pillared beidellite (APB) prepared from aluminum chlorohydrate.

where S is the specific surface area; σ is the thickness of a single molecular layer of N_2 molecules (σ = 3.54 Å); K is a constant equaling 4.37 m^2/cm^3; and b_{t_0} is the slope of the line passing through the origin of the t-plot.

A second domain of data points below t = 6 is fitted to another straight line. The intercept of this line gives the microporous STP volume (V_{STP}), which can then be converted to a microporous liquid volume (V_{liq} = 0.00154 V_{STP}) and to an equivalent microporous surface area ($S_{mic} = KV_{STP}$). The slope of this second line (b_{t_1}) is the mesopore plus macropore surface area [$S_{(meso + macro)} = \sigma K b_{t_1}$].

Micropore sizes in zeolitic materials may be estimated by measuring the adsorption of probe molecules of various kinetic diameters[7, 15, 22] This method can also provide information on diffusion limitations in the catalysts.[23] Pillared clays usually exhibit a broader pore size distribution than zeolites. With reliable data on argon or nitrogen adsorption in the low pressure range, the micropore size distribution can be estimated from the isotherm.[24] Vaughan[25] reported a micropore distribution centered near 9–10 Å for an alumina pillared clay prepared from a Chlorohydrol® pillaring agent. The value obtained is close to the gallery height measured using XRD, but it may not actually relate to the lateral interpillar spacing.

The desorption branch of an adsorption–desorption isotherm is generally favored by researchers for evaluating the mesopore size distribution. A slit-shaped or parallel pore model is appropriate for pillared clays. The mesopore and external surface areas obtained using this method should be in agreement with t-plot results.

There is a growing interest in the use of ^{129}Xe NMR for characterizing the micropores of solid acids. Fetter et al.[26] have reported that the apparent pore size (about 10 Å) of alumina pillared clays is near the gallery height measured using XRD.

Surface Acidity

Temperature-programmed ammonia desorption is a simple technique that can be used to determine the acidity of solid acids.[27] Ammonia is first adsorbed on a sample at about 100 °C, then the temperature is raised and the additional amount of ammonia desorbed is recorded by a thermal conductivity detector (TCD) cell. However, layer dehydroxylation of pillared clays can occur between 450 and 500 °C, causing the peak for desorbed water to be superimposed upon the desorption peak for strongly adsorbed ammonia. Thus, preheating the sample to the dehydroxylation temperature might modify the acidity of the solid. Matsuda et al.[28] overcame this problem by running a blank experiment for each sample, subtracting the results from the temperature-programmed desorption (TPD) spectrum. An alternative solution is to trap the desorbed ammonia in H_2SO_4 and then titrate.[29] Unfortunately, both these methods can produce ambiguous results, and there is some disagreement among researchers concerning acidity measurements taken using them. Moreover, NH_3 is a small molecule that could access sites inaccessible to bulkier organic molecules.

Thermodesorption of pyridine observed using IR spectroscopy is a powerful method for studying the surface acidity of solids because it allows Lewis and Brönsted acidities to be distinguished.[30] Pyridine adsorbed on Lewis acid sites exhibits absorption bands at 1450 and 1490 cm^{-1}, whereas Brönsted site absorption occurs at 1490 and 1540 cm^{-1}. Acid strength is estimated by the desorption temperature of the probe. The kind and strength of acidity depends on the clay layer. For example, Poncelet and Schutz[31] showed that pillared beidellite has strong Brönsted acidity, whereas pillared montmorillonite exhibits only Lewis acidity. They also observed a correlation between the acidity and the intensity of the hydroxyl group IR absorption band.

^{29}Si and ^{27}Al MAS-NMR

The growing use of MAS-NMR characterization over the last 15 years has provided a better comprehension of the structure of minerals[32, 33] As noted by Fripiat[34] in his excellent review of the topic, MAS-NMR is sensitive to short-range order, making it especially useful for characterizing poorly crystallized materials such as pillared clays. Owing to MAS-NMR examinations of ^{29}Si and ^{27}Al nuclei, the nature of the pillaring species and their interaction with the clay layer have become understood.[35] Studies of the transformations that occur during calcination—for example, crosslinking the pillar to the layer in beidellite[35] and fluorohectorite[36] or proton migration from the pillars[37]—are also greatly facilitated through the use of MAS-NMR.

   ALUMINA PILLARED CLAYS...   Chapter 8

There are inherent difficulties associated with the MAS-NMR of clay minerals. For one, ^{29}Si MAS-NMR is often preferred over ^{27}Al MAS-NMR because the nuclear electric quadrupole interaction of the latter can produce very broad NMR peaks for powder samples.[32] Also, studies have often been restricted to clays with low iron content because of the line-broadening due to the relaxation effects of paramagnetic centers.[38]

^{29}Si chemical shifts are primarily affected by the nature of the neighboring atoms in its oxygen tetrahedron and by the Si–O–Si bond angle. In phyllosilicates, where each Si site is linked to three other tetrahedral sites (i.e., Q^3 environments), Si atoms with no aluminum nearest-neighbor atoms in the tetrahedral layer exhibit shifts between –99 and –93 ppm. When one of the nearest-neighbor atoms is aluminum, the resonance appears between –85 and –90 ppm. This chemical shift difference permits the differentiation of beidellite from most other clays, since aluminum is present in its tetrahedral layer. Tennakoon et al.[37] have demonstrated a relationship between the ^{29}Si chemical shift and the clay layer charge, attributing the change in chemical shift toward more negative values (higher field) after the calcination of pillared clay to a lowering of layer charge, which is caused by proton migration from the gallery to the layer.

The distinction between tetrahedrally and octahedrally coordinated aluminum is clearly shown using ^{27}Al MAS-NMR. Octahedral aluminum appears between 0 and 10 ppm, whereas tetrahedral aluminum appears between 60 and 70 ppm. However, relative intensities are very difficult to analyze, because the peak shape is very sensitive to the symmetry of the coordination shell. The tetrahedra aluminum atom at the center of the Al_{13} oligomer is observed[35] at 62.3 ppm in uncalcined pillared beidellite (see Figure 8.7). This strongly supports the hypothesis that Al_{13} is indeed the intercalated species.

Pillared Clays As Catalysts

The study and development of pillared clays has been greatly encouraged by the need for new cracking catalysts (FCC) capable of cracking heavy feedstocks.[39] Because the pores of pillared clays are larger than those of zeolites, pillared clays seemed to be well-suited for the refining of heavy oil fractions. The first micro-activity tests[2, 40] (MAT) revealed two shortcomings of pillared clays: (1) their hydrothermal stability, and thus their regenerability, was poor compared to zeolites and (2) coke formation was at least two times higher for pillared clays than for zeolites. Since that time, the thermal stability of pillared clays have been much improved largely through the efforts of Guan and co-workers.[41] These researchers were the first to pillar rectorite, an interstratified, layered silicate, in which the layer stacking consists of a regularly alternating sequence of mica-like and montmorillonite-like layers. Pillared rectorite is stable under hydrothermal conditions up to 800 °C.

Several studies have been done to reveal the cause of high coke formation on pillared clays. The iron present in natural clays has been suggested as a major

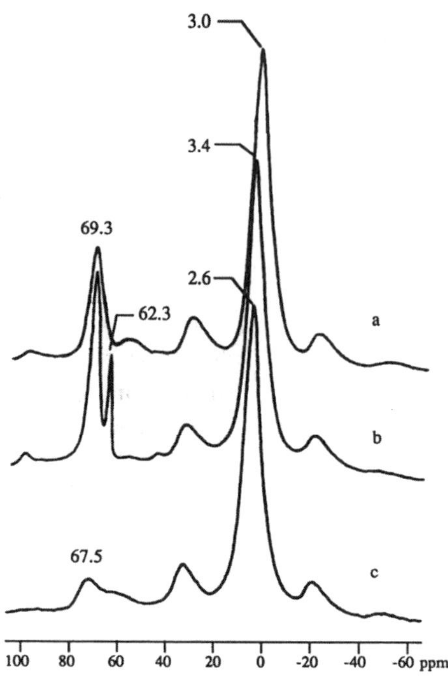

Figure 8.7 ^{27}Al MAS-NMR spectra of (a) beidellite; (b) uncalcined pillared beidellite; and (c) calcined pillared beidellite. (Adapted from Plee et al.[35])

catalyst for coke formation. A recent study by Occelli et al.[42] shows that a determining factor is the localization of the iron in the layer between the clay particles or in the pillars. Iron far from the acid centers does not increase coke formation. The high coke yield may also be related to the relatively large micropores of pillared clays in which the polymerization of hydrocarbons can occur. A careful characterization of the pore structure of pillared clays coked after biphenyl alkylation showed that coke formation did indeed occur in the micropores.[43]

Pillared clays are a versatile class of acidic materials possessing micropores that can be modified to a wide range of sizes. Thus, it is of interest to study shape-selective effects on reactions with these catalysts. Pillar height can be modified by varying the hydrolysis ratio[44] of Al^{3+} or by incorporating other elements (e.g., La, Ce, or Si) into the pillars.[45] Matsuda et al.[44] observed a relationship between pillar height and shape-selective production of 1,2,4,5-tetramethylbenzene by the disproportionation of 1,2,4-trimethylbenzene. An interesting method for modifying pore size is to vary the pillar density through strategies that include modification of the layer charge density by charge reduction methods,[46] re-intercalation of the pillared clay after ammonia treatment,[47] and using clays of different charge densities.[43] However, varying the clay can give rise to pillared clays with very different pore structures, which can have significant effects on catalytic properties. For example,

   ALUMINA PILLARED CLAYS...   Chapter 8

as was shown earlier, a clay of small particle size like the synthetic hectorite Laponite yields an Al[13] reaction product that is much less microporous than pillared montmorillonite. Matsuda et al.[48] observed a modified product selectivity for the gas phase disproportionation of 1,2,4-trimethylbenzene over the microporous pillared montmorillonite or the less microporous pillared saponite. The mesopore structure of pillared clays can affect not only activity, but also product distribution.[43] The propene alkylation of liquid phase biphenyl results in more than 30% orthoalkylated product when non-microporous delaminated Laponite is used as the catalyst in contrast to only 15% orthoalkylation with pillared montmorillonite at the same conversion rate. The dependence of reactivity and selectivity on pillared-clay particle size demonstrates that liquid phase alkylation is under strong diffusion control.

8.4 Summary

Alumina pillared clays are promising catalytic materials that can be synthesized using a wide range of host clays and pillaring reagents. In the case of alumina pillared clays, derivatives can be prepared by the reaction of the Na^+-exchange form of the smectite precursor with either base-hydrolyzed Al^{3+} or a commercially available aluminum chlorohydrate solution that contains the derived Al_{13} oligomers. These two pillaring reagents afford compositionally distinguishable products.

X-ray diffraction and surface area measurements are essential for the characterization of alumina pillared clays. XRD reveals whether an intercalation reaction has ocurred between the clay layers, and surface measurements demonstrate that the intercalated pillars are sufficiently spaced laterally to generate an internal microporous volume. With additional refinements, these methods can provide more information on the crystallinity and on the pore structure of the pillared clay. In order to understand the catalytic properties of pillared clays, it is very important to characterize their surface acidity by well known acid–base techniques. ^{29}Si and ^{27}Al MAS-NMR can assist in the interpretation of these acid–base properties.

Acknowledgments

We are grateful to the National Science Foundation for financial support of this work through grant DMR-8903579. J.-R. Butruille wishes to thank Rhône-Poulenc for a graduate fellowship in support of his research.

References

1   R. Burch, Ed. "Pillared Clays," In *Catal. Today.* **2** (2,3), 1988.

2   F. Figueras. *Catal. Rev.-Sci. Eng.* **30** (3), 457, 1988.

3   T. J. Pinnavaia and H. Kim. In *Zeolite Microporous Solids: Synthesis, Structure and Reactivity.* (E. G. Derouane, Ed.) Kluwer, Belgium, forthcoming.

4   R. M. Barrer and D. M. MacLeod. *Trans. Faraday Soc.* 1290, 1955.

5   G. W. Brindley and R. E. Sempels. *Clay Miner.* **12**, 229, 1977.

6   N. Lahav, U. Shani, and J. Shabtai. *Clays Clay Miner.* **26**, 107, 1978.

7   D. E. W. Vaughan and R. J. Lussier. In *Proc. 5th Int. Conf. Zeolites.* (Rees and V. C. Lovat, Eds.) Heyden, London, 1980, p. 94.

8   S. Yamanaka and G. W. Brindley. *Clays Clay Miner.* **27**, 119, 1979.

9   T. J. Pinnavaia, M. Tzou, and S. D. Landau. *J. Am. Chem. Soc.* **107**, 2783, 1985.

10   S. Yamanaka and M. Hattori. *Catalysis Today.* **2**, 261, 1988.

11   E. G. Rightor, M.-S. Tzou, and T. J. Pinnavaia. *J. Catal.* **130**, 29, 1991.

12   J. Sterte. *Clays Clay Miner.* **35**, 658, 1986.

13   A. Schutz, W. E. E. Stone, G. Poncelet, and J. J. Fripiat. *Clays Clay Miner.* **35**, 251, 1987.

14   J. R. Harris. In *Perspectives in Molecular Sieves Science.* (W. H. Flank and T. E. Whyte, Eds.) American Chemical Society, Washington, D.C., 1988, p. 253.

15   T. J. Pinnavaia, M.-S. Tzou, S. D. Landau, and R. H. Raythatha. *J. Mol. Catal.* **27**, 195, 1984.

16   D. M. Moore and R. C. Reynolds, Jr. *X-ray Diffraction and the Identification and Analysis of Clay Minerals.* Oxford University Press, New York, 1989.

17   M. L. Occelli, S. D. Landau, and T. J. Pinnavaia. *J. Catal.* **104**, 331, 1987.

18   A. C. D. Newman and G. Brown. In *Chemistry of Clays and Clay Minerals.* Wiley, New York, 1987.

19   J. Shabtai, M. Rosell, and M. Tokarz. *Clays Clay Miner.* **32**, 99, 1984.

20   M. L. Occelli and J. Lynch. *J. Catal.* **107**, 557, 1987.

21   S. J. Gregg and K. S. W. Sing. *Adsorption, Surface Area and Porosity.* 2nd ed., Academic Press, London, 1982.

22   M. L. Occelli, V. N. Parulekar, and J. W. Hightower. In *Proc. 8th Inter. Congress Catalysis, Vol. 4.* Verlay Chemie, Weinheim, Germany, 1984, p. 725.

23   V. N. Parulekar and J. W. Hightower. *Applied Catalysis.* **35**, 249, 1987.

24   G. Horvath and K. Kawazoe. *J. Chem. Eng. Japan.* **16**, 470, 1983.

25   D. E. W. Vaughan. In *Perspectives in Molecular Sieve Science.* (W. H. Flank and T. E. Whyte, Eds.) American Chemical Society, Washington, D.C., 1988, p. 308.

26   G. Fetter, D. Tichit, L. C. De Ménorval, and F. Figueras. *Applied Catalysis.* **65**, L1, 1990.

27 B. M. Lok, B. K. Marcus, and C. L. Angell. *Zeolites.* **6**, 185, 1986.

28 T. Matsuda, M. Asanuma, and E. Kikuchi. *Applied Catalysis.* **38**, 289, 1988.

29 M. Kojima, R. Hartford, and C. T. O'Connor. *J. Catal.* **128**, 487, 1991.

30 E. P. Parry. *J. Catal.* **2**, 371, 1969.

31 G. Poncelet and A. Schutz. In *Chemical Reactions in Organic and Inorganic Constrained Systems*, NATO ASI Series C. (R. Setton, Ed.) D. Reidel Publishing Co., Dordrecht, Holland, 1986, p. 165.

32 C. A. Weiss, Jr., S. P. Altaner, and R. J. Kirkpatrick. *American Mineralogist.* **72**, 935, 1987.

33 D. E. Woessner. *American Mineralogist.* **74**, 203, 1989.

34 J. J. Fripiat. *Catalysis Today.* **2**, 281, 1988.

35 D. Plee, F. Borg, L. Gatineau, and J. J. Fripiat. *J. Am. Chem. Soc.* **107**, 2362, 1985.

36 T. J. Pinnavaia, S. D. Landau, M.-S. Tzou, I. D. Johnson, and M. Lipsicas. *J. Am. Chem. Soc.* **107**, 7222, 1985.

37 D. T. B. Tennakoon, W. Jones, and J. M. Thomas. *J. Chem. Soc, Faraday Trans. I.* **82**, 3081, 1986.

38 D. T. B. Tennakoon, J. M. Thomas, W. Jones, T. A. Carpenter, and S. Ramdas. *J. Chem. Soc., Faraday Trans. I.* **82**, 545, 1986.

39 D. E. W. Vaughan. *Catalysis Today.* **2**, 187, 1988.

40 R. J. Lussier, J. S. Magee, and D. E. W. Vaughan. In *7th Canadian Symp. Catal., Prepr.* Chem. Inst. Canada, Ottawa, 1980, p. 88.

41 J. Guan, E. Min, and Z. Yu. *Proc. 9th Int. Conf. Catal.* (M. J. Philips and M. Ternam, Eds.) Chem. Inst. Canada, Ottawa, 1988.

42 M. L. Occelli, J. M. Stencel, and S. L. Suib. *J. Mol. Catal.* **64**, 221, 1991.

43 J.-R. Butruille and T. J. Pinnavaia. *Catalysis Today.* Forthcoming.

44 E. Kikuchi, T. Matsuda, H. Fujiki, and Y. Morita. *Applied Catalysis.* **11**, 331, 1984.

45 J. Sterte. *Clays Clay Miner.* **39**, 167, 1991.

46 T. Mori and K. Suzuki. *Chem. Letters.* 2165, 1989.

47 E. Kikuchi, H. Seki, and T. Matsuda. *Stud. Surf. Sci. Catal.* **63**, 311, 1991.

48 T. Matsuda, H. Nagashima, and E. Kikuchi. *Applied Catalysis.* **45**, 171, 1988.

Appendix: Technique Summaries

The technique summaries in the following pages of this appendix include many one page summaries marked with an asterisk. These techniques are fully described in the *Encyclopedia of Materials Characterization* by C. Richard Brundle, Charles A. Evans, Jr., and Shaun Wilson. The remaining summaries are for techniques which do not appear in that volume.

Auger Electron Spectroscopy (AES)* **1**

In Auger Electron Spectroscopy (AES), a focused beam of electrons (typically a few keV to a few 10's of keV in energy) strikes the sample, causing electrons to be ejected. Some of these (the Auger electrons, named after Pierre Auger who first observed them) have kinetic energies characteristic of the atoms from which they came and so identification of the elements present in the sample can be made directly from the measurement of these energies. On a finer scale, it is sometimes also possible to determine the chemical state of the element from small shifts in the Auger energies caused by the chemical bonding of the element (chemical shift). For a solid, AES probes 2–20 atomic layers deep, depending primarily on the kinetic energy of the ejected Auger electron concerned, and somewhat on the material. With appropriate standards the relative concentration of the elements present within the probing depth can be estimated from the relative intensities of the Auger peaks in the spectrum. The great strength of AES is the high spatial resolution achievable through use of a focused beam (down to 10's of nm), and the ability to combine measurement with ion sputtering material removal, to obtain a three-dimensional elemental profile. The main use is with metals and conducting or semiconducting inorganic materials, since beam damage and charging can be issues with non-conducting and organic materials. Many Auger spectrometers are designed to maximize sensitivity (typically down to 100 ppm of element concentration) and speed of analysis, and do not have the spectral resolution to use chemical shifts for chemical state identification. Auger is capable of identifying all elements, except hydrogen and helium.

2

In Secondary Ion Mass Spectrometry, SIMS, the sample is bombarded by a primary ion beam (typically in the few 100 ev to many keV energy range). Several alternative ion species are used, depending on the requirements of the analysis, which causes atoms and clusters of atoms to be sputtered from the sample surface. These atoms and clusters consist largely of neutral species, but a small fraction of both positively and negatively charged species occur (the secondary ions). These secondary ions are passed into a mass spectromter (several different types are in use), where the number of ions at each mass/charge ratio are counted. The secondary ion counts may be turned into atomic concentartions present in the sample, if (and only if), comparison to standards of the same or very similar material is available. In Dynamic SIMS, the secondary ion signal intensities are monitored as a function of sputter time (depth), allowing concentrations to be determined as a function of depth.

The strength of Dynamic SIMS is the ability to to monitor for all atoms, including hydrogen, helium, and including isotopes, over a wide dynamic range of concentrations (down to ppb to ppt levels). A lateral resolution down to below 10 nm is possible, depending on the nature of the instrument and sample (not at trace ppb levels). A depth resolution of 2–30 nm is possible, depending on material, sputter conditions, and the depth sputtered. The first few nm depth sputtered, however, is rarely quantifiable, however, owing to the initial non-equilibrium nature of the sputtering process and the lack of appropriate standards for a region which can be substantially different from the bulk material.

The major use of Dynamic SIMS is the quantitative determination of the concentration of dopants in semiconductor material, as a function of depth, over a dynamic range of 5 orders of magnitude. It is also used in a similar way for trace impurities in metals and alloys and geological materials.

If an energy analyzer is added to a Transmission Electron Microscope (TEM) or Scanning Transmission Electron Microscope (STEM) after the image plane, the energy losses suffered by the incident electron beam as it passes through the thin TEM/STEM sample can be measured. The losses are caused by plasmon excitation (a few 10's eV energy) and atomic core level excitation. Atomic core binding energies are characteristic of the atom concerned, so the threshold energy for core level excitation provides a direct atomic identification (cf. XPS). Relative concentrations of elements present in the sample can be determined from the relative intensities of the energy loss threshold. Near-edge structure on the energy loss features can also provide bonding information (cf. XPS). The sample must be very thin (100 Å to 2000 Å, depending on beam energy) to avoid multiple inelastic scattering, resulting in peak broadening, and loss of intensity. A consequence of avoiding multiple scattering is a lack of beam spreading as the electrons pass through the sample, which results in the lateral resolution being determined largely by the width of the incident beam. In STEM mode a lateral resolution of a few Å is therefore possible. The main use of EELS is for elemental and chemical composition analysis at ultra-high spatial resolution (at grain boundaries, for instance).

The electron paramagnetic resonance (EPR), or electron spin resonance (ESR), method can characterize paramagnetic ions and radicals because of its ability to detect unpaired electrons. In EPR experiments, a solid sample—crystalline or powder—is placed in an external magnetic field that splits the energy levels (allowed spin states) of the unpaired electron. EPR can detect the presence of a paramagnetic species and its oxidation state and provide a detailed description of the bonding and orientation of a surface complex. The limitation of the technique is that it is applicable only to isolated paramagnetic species.

Main use	Detection of paramagnetic species; oxidation states of transition metal ions; bonding and orientation of surface complex
Unique features	Highly sensitive; in situ
Destructive	No
Elements detected	All elements possessing an unpaired electron
Detection limit	10^{11} spins
Depth examined	Representative of bulk; provides surface information for adsorbed species
Typical sample size	10 mg to several grams
Measurements conditions	Usually at ambient pressure and temperature
Sample form	Powder or single crystal
Instrument cost	$100,000
Space requirement	100 ft^2

In Electron Microprobe X-Ray Microanalysis (EPMA) a beam of electrons (5–30 keV) is focused onto a solid sample and the X rays emitted (fluorescence) are analyzed for their energies and intensities. The energies are characteristic of the atoms undergoing the emission, providing an atomic identification (Be to the actinides), and the intensities can be related to the concentrations present, providing a quantitative analysis. The X rays are detected and measured using either an Energy Dispersive X-Ray Spectrometer (EDS), or a Wavelength Dispersive X-Ray Spectrometer (WDS). EPMA systems usually have both methods available (EDS is experimentally simpler and faster; WDS is more accurate and can reach lower detection levels), together with SEM capability. For smooth, flat samples, detailed correction procedures for matrix effects (backscattering which varies with atomic number; secondary X-ray absorption and fluorescence occurring during emission), together with elemental standards, allow quantification typically to about +/– 4%. For rough surfaces, or particles, analysis is usually only semi-quantitative. The detection limit for EDS is about 1000 ppm and about 100 ppm for WDS. Probing depths, and spatial resolution, depend on the incident beam energy, ranging from about 0.1 μm at the lowest energy to 5 μm at high energy. Quantitative mapping on flat surfaces is common. The major use of EPMA is nondestructive quantitative elemental analysis of major and minor constituents in the top few μm of inorganic materials.

Energy-Dispersive X-Ray Spectroscopy, EDS, is a specific technique for the detection and energy distribution determination of X-Ray Fluorescence, XRF. XRF is the phenomena where X rays are emitted from a material when bombarded by high energy radiation (electrons, ions, X rays, neutrons, gamma rays). Some of the X-ray energies emitted are characteristic of the atoms present, allowing atomic identification in the material of interest.

In EDS a solid state X-ray detector, usually lithium drifted Si, and pulse counting electronics are used. The detector converts an incoming X-ray photon into an electronic pulse of amplitude proportional to the energy of the X ray. The signal processing electronics counts the number of pulses of each different amplitude, resulting in a histogram of X-ray energy versus intensity. The X-ray energies allow atom identification, and the relative peak intensities can be related to relative atomic concentrations by comparision to standards, or by theoretical calculations.

All elements with an atomic number higher than Li are, in principle detectable, though effectively dealing with low Z elements requires some care and correctly set up instrumentation. In practise, EDS is primarily used in conjunction with e-beam columns (SEMs, TEMs, STEMs, Auger instruments) as the excitation source. The depth probed is dependant primarily on the energy of the electron beam, and on the material being probed. It can vary from as little as 20 nm (high Z material, low beam energy) to as much as 5000 nm (low Z material, high beam energy). The lateral resolution is similar to the depth probed but is not determined by the primary beam diameter. It is determined by the beam energy and the sample material, because the beam spreads by scattering as it penetrates into the material, creating X rays in this enlarged volume.

The particular strength of EDS is the simultaneous (or parallel) detection of elements rapidly and cheaply, by placing the physically small detector inside the SEM, or other electron beam system. As such it adds an elemental analytical capability to imaging electron beam columns. Its drawback is that the typical solid state detector has very poor energy resolution, which means that there are significant peak overlaps, sometimes making it difficult to distinguish elements. This is particularly true in the low energy region.

Extended X-Ray Absorption Fine Structure (EXAFS)*   7

Extended X-Ray Absorption Fine Structure (EXAFS) requires a tunable source of monochromatic X-rays, which is nearly always supplied using a national synchrotron radiation facility. The photon beam impinges on the sample (either solid, liquid, or gaseous), and the photon energy is scanned from below a core-level B.E. of a selected element to well above it, while the X-ray absorption is monitored as a function of the energy in one of several ways (directly, in transmission; from emitted secondary electrons, in reflection; or by X-ray fluorescence, either in reflection or transmission). If the selected element is present in the sample, it will be identified by a sharp increase in absorption at the core level B.E. The intensity of the absorption is related to the concentration of that element. Beyond the absorption edge, oscillations in absorption occur as the incident photon energy increases, the EXAFS structure. The frequencies of the oscillations are inversely proportional to the interatomic spacings between the absorbing atom and its nearest neighbors (destructive and constructive interference, sensitive up to about 5Å distance). The amplitude is related to the type and number of neighboring atoms. Providing accurate structure solutions from the data is, however, heavily dependant on the use of model compounds with known structures. In addition to the oscillating EXAFS structure, there is structure on the near edge of the absorption (Near Edge X-Ray Absorption Fine Structure, NEXAFS) which carries chemical state information. The depth probed in a solid sample depends upon which detection scheme is used and can be varied from Å's to µm's. The main use of EXAFS is to provide element specific information on the local atomic environment, particularly in situ for catalysts and electrochemical systems. Its main drawback is the need to work at a synchrotron facility.

In Infrared Spectroscopy (IR), a beam of light (electromagnetic radiation) in the infrared wavelength region impinges on the sample and the wavelength/frequency is scanned. Whenever there is a match of the light frequency to the vibrational frequency of a vibrational mode within the sample (i.e., vibration of atomic positions) it is possible that the radiation at that frequency will be absorbed and the vibration excited (selection rules apply and not all vibrations can be excited). The absorption occurring, as a function of IR wavelength, is monitored by comparing the input intensity of the radiation to the output intensity, thus revealing the vibrational frequencies existing in the sample. Since a vibrational frequency value is dependant on the bonding between the atoms involved in that vibration, vibrational frequencies of materials are characteristic of the chemical groups existing in the material (e.g., OH, CN, CH_2, COOH, etc.), so determination of the vibrational frequencies present allows determination of which chemical groups are present. With standards the concentrations of the groups present can also be determined from the strength of the absorptions.

Fourier Transform IR simply describes the most common modern technique for scanning the wavelength of the impinging IR beam. It is based on the Michelson interferometer, where constructive and destructive interferences between the two halves of a split light beam are controlled by changing the path length of one of the beams with respect to the other by moving reflecting mirrors. Since the path length difference for destructive/constructive interference depends on wavelength, the oscillation of the mirrors provides a way of scanning the wavelengths from a polychromatic light source.

For solids, IR spectroscopy can be performed either in reflection or in transmission if the sample is thin enough to pass the radiation. This varies enormously with material because absorption coefficients vary enormously (some materials are transparent to IR, some highly absorbing). In reflection, coupling the radiation to an optical microscope allows a lateral resolution down to 20 µm, with the depth probed depending on absorption coefficient, but usually being many µm. Another reflection mode is Attenuated Total Reflection, ATR, where light enters at grazing angle and probes only the surface region (10's of nm). IR of solids does not require a vacuum, but the path length of the radiation through ambient atmosphere must be purged using dry nitrogen, to avoid gas phase absorption from strong absorbers such as water vapor.

The strength of IR is the qualitative and sometimes quantitative (with standards) identification of the presence of chemical species, or functional groups, sometimes down to trace levels in liquids and solids. For solids stress, strain, crystallinity/ amorphousity, and inhomogeneities can be detected from absorption peak broadenings and shifts.

High Resolution Electron Energy Loss Spectroscopy (HREELS)* 9

In High Resolution Electron Energy Loss Spectroscopy (HREELS), a beam of low energy electrons (1–10 eV) which has been passed through a high resolution electrostatic electron monochromator so that the beam spread is only a few meV, is directed onto a sample and the back-scattered electrons exiting the surface at a quasi-specular angle are analyzed for their energy and intensity distribution. The analysis is usually performed with an electrostatic analyzer identical to the monochromator. A fraction of the electrons incident on the sample surface cause vibrational excitations in the sample, leading to energy losses in the scattered beam equal to the vibrational energies concerned. A vibrational spectrum of the material is thus obtained by measuring these energy losses. It is equivalent to the vibrational spectrum obtained by Raman scattering or IR absorption, but is confined much closer to the surface (a few nm). It also has a more restricted vibrational frequency range and much poorer spectral resolution. The major use of HREELS is to identify chemical functional groups present by monolayer adsorption on well-defined single crystal metal or semiconductor surfaces, under ultra-high vacuum conditions. As such it is really a research technique. It can also be used on flat thin film surfaces, such as polymers. There is basically no spatial resolution capability (mm size) and no easy quantification.

Inductively Coupled Plasma Mass Spectrometry (ICPMS)* **10**

In Inductively Coupled Plasma Mass Spectrometry (ICPMS), the sample is introduced by one of several ways into an inductively coupled plasma which fragments and ionizes all species present down to the atomic level. The ions are passed into a mass spectrometer (several types are in use) and mass analyzed to reveal which atoms are present. The plasma is highly efficient, and, coupled with an efficient and high resolution mass spectrometer, detection limits down to sub ppb are achievable for many elements. If a low resolution mass spectrometer is used, mass interferences limit the detection limit for some elements, notably Fe. Rough quantification to the 5 to 20% level is easy using standard sensitivity factors determined for the mass spectrometer. For better quantification standard solutions using a material matrix as similar to the sample as possible are used. Sub 1% accuracy can be achieved by using the method of isotope spiking.

The normal method of sample introduction is to nebulize an aqueous solution into the plasma. For solids, dissolution by a solvent, or digestion by an acid (normally nitric acid), is the preferred method, but direct sampling using laser ablation is possible. This method is the only way to achieve any spatial resolution. Laser ablated craters are typically 10's of μm wide and a fraction to a few μm deep.

The main uses of ICPMS are in two areas. One is the routine determination of contaminant levels in solutions and solvent, particularly water supplies. The second is for determination of trace levels of elements in semiconductors and alloys.

Inductively Coupled Plasma-Optical Emission Spectroscopy (ICP-OES)*   **11**

In Inductively Coupled Plasma Optical Emission Spectroscopy (ICP-OES), a solid sample is first dissolved into aqueous solution and then nebulised and passed, in aerosol form, into an inductively coupled plasma region. The plasma atomizes and ionizes the species present, leaving them in electronically excited states. Decay from these states results in optical emission in the visible and UV regions. The wavelengths of the emissions are characteristic of the atoms concerned, providing atomic identification, and the intensities are proportional to the concentrations present. The wavelengths are analyzed by a grating spectrometer (or several spectrometers). Quantitation is achieved using aqueous solution standards and calibration. Not all elements can be detected, since about one third of the elements in the periodic table do not produce suitable emission lines. ICP-OES is a trace analysis method, with typical sensitivities ranging from sub-ppb to 100 ppb, with typically 10% accuracy. Better than 1% accuracy can be achieved using specialized calibration methods. In addition to trace analysis for bulk solids, liquids can be analyzed directly, and material at surfaces or in thin films can be introduced directly into the plasma by special methods such as laser ablation. Laser ablation also allows some spatial resolution (10's µm), though controlling this is difficult.

ICP-OES is mainly used for inexpensive, rapid, semi-quantitative measurement of trace and minor elements in solids and liquids. For better trace level sensitivity and for elements not accessible by ICP-OES, ICP-Mass Spectrometry can be used as a more expensive alternative.

In Ion Scattering Spectroscopy (ISS), a low-energy, monoenergetic beam (100 to 5000 eV) of ions of an inert element (usually He^+, Ne^+, or Ar^+) is directed onto a solid surface, and the energy of the backscattered ions is measured at a fixed angle. The collision is approximately kinematic, following the simple laws of momentum transfer for a binary collision between atoms of different mass. The energy loss of the primary beam, therefore, identifies the mass of the atom doing the backscattering. The energy loss spectrum is measured by passing the ions around an electrostatic energy analyzer and is usually plotted as the relative intensity of ions versus relative energy, E/E_0, where E is the energy of the backscattered ion, and E_0 is its original incident energy. This plot can be converted directly into approximate concentration versus atomic number, Z. The sharp discrete peaks in the spectrum are produced only by collisions with atoms in the outermost atomic layer, making the technique extremely surface sensitive. At these energies, ion sputtering occurs as the data is being collected, so unless the total fluence is kept low, surface damage occurs. A depth profile over 100 Å can be achieved by measuring during deliberate sputtering. The major use of the technique is to provide exclusive detection in the outer monolayer of material. The major drawback is poor spectral resolution, which limits the ability to distinguish many common adjacent elements in the periodic table. This can be partly overcome by the correct choice of probe ion and scattering angle for the element concerned. Detection limits go roughly as Z^2 and can be as good as 50 ppm for heavy elements in light matrices, or as poor as a few percent for low Z elements such as C.

Low-Energy Electron Diffraction (LEED) is a technique for providing a two-dimensional diffraction pattern from a surface. It is only applicable to a surface with long-range order, typically a single crystal surface, or an adsorbed over layer. The method involves impacting a collimated beam (typically 0.1 mm diameter) of monoenergetic electrons, in the energy range 10 to 1000 ev (equivalent to a photon wavelength of 0.4 to 4 Å) onto the surface at normal incidence. Electrons in this energy range have very short inelastic mean free path lengths, so the electrons which are elastically diffracted back from the sample only come from the first few Å depth. They are separated from inelastically scattered electrons, which have lost energy, by a retarding grid system and are then detected on a phosphor screen. The observed diffraction pattern reveals the 2D unit cell type and dimensions. Determination of the actual surface atomic atom positions relative to each other, or the underlying layers, requires measurement of the diffraction intensities as a function of varying the electron energy and comparison to theoretical calculations on trial structures. The shapes, splittings, and broadening of the diffracted beams also carries information on surface disorder.

The major use of LEED has been to determine surface structures of both clean bulk crystalline material and adsorbed layers on such surfaces, by comparison to theory. It is also used routinely and simply to monitor the surface cleanliness of single crystal surfaces, and how well ordered the surface is (sensitive over about a 200 Å range in most instruments). A very specialized version exists, the Low Energy Electron Microscope, LEEM, which functions just like an imaging TEM and has a surface imaging resolution of 150 Å.

The Mössbauer effect is the result of the recoil-free emission and resonant reabsorption of low energy γ rays by atoms in solids. The Mössbauer apparatus consists of an emitter, an absorber, and a γ-ray detector. For the Mössbauer effect to occur, the nuclear transitions must be recoil-free (the recoil momentum must be transmitted to the solid as a whole), with no new phonons excited in the solid. The chemical shift of a Mössbauer line depends on the electronic charge density at the nucleus; this originates mainly from s electrons. Additional changes may also be produced indirectly through screening effects on s electrons by p and d electrons. The very narrow line width of Mössbauer γ radiation allows very small perturbations in the sample environment to be measured.

Main use	Phase identification and quantification; structural characterization of disordered states
Unique features	Element-selective; in situ
Destructive	No
Elements detected	Elements possessing proper radioisotope-emitting γ rays (Fe, Co, Sn, and I)
Detection limit	$\sim 10^{18}$ atoms of the nuclear isotope studied
Depth examined	Representative of bulk; surface information for highly dispersed systems
Typical sample size	500 mg to several grams
Measurement conditions	Usually at ambient temperature and pressure
Sample form	Powder or single crystal
Instrument cost	$25,000
Space requirement	10 ft^2

Neutron Activation Analysis (NAA) is a technique for trace analysis in bulk material. It involves two separate steps. The sample is initially irradiated by neutrons in a nuclear reactor to create the radio isotopes (which may take 1 to 7 days). This is followed by removal to a gamma-ray spectroscopy facility, where the gamma-rays given off during subsequent decay of the radio isotopes can be monitored. The gamma-ray energies emitted are characteristic of the atoms decaying and are determined using a solid state energy dispersive detector in an analogous manner to the X-ray analysis in EDS (see appendix 7), except that the energies are in the MeV range. The technique is a bulk one, since both neutrons and gamma-rays penetrate deeply. There is also no spatial resolution.

Both the initial neutron capture probability (cross-section) to create the radio isotopes, and the half-lives for the subsequent decay process vary enormously, in no particular pattern, across the periodic table. In fact, only about two thirds of the elements can produce radio isotopes this way. So trace element sensitivity varies enormously among elements, and also with the matrix, which may give off competing gamma-rays. For trace metals in Si or SiO_2, ppb to ppt sensitivity is easily obtained, but other semiconductor matrices such as GaAs are less favorable. Trace elements in plastics and biological materials can be measured because the matrix elements of C, O, H and N all have low neutron capture cross-sections. The material may suffer radiation and heating damage in the process, however.

The major uses of NAA are for trace determination of bulk impurities in semiconductor, biological, geological, and envonmental samples, and in forensic science. Quantification can be achieved from signal intensities using standards, or from calculations if enough information about the instrumetal conditions is available. The major drawbacks are the need to use reactor and certified gamma-ray facilities, the long time involved for an analysis, and the special handling needed for the samples.

Neutron Diffraction uses thermal neutrons of wavelength, λ, 1–2 Å to determine the atomic structure in crystalline phases in an essentially similar manner to X-Ray Diffraction. Neutrons are diffracted according to Bragg's Law ($\lambda = 2d \sin\theta$, where d is the spacing between atomic planes). The intensity of the diffracted beams is measured as a function of either the diffraction angle, 2θ, or by scanning λ for a fixed θ. The former is done using neutrons of a fixed, known wavelength from a nuclear reactor. The later involves using a high energy beam of photons impinging on a heavy metal, which produces pulses of neutrons (a spallation source). The major difference between Neutron Diffraction and X-Ray diffraction is (a) the neutrons penetrate solids much deeper, so that the analysis is always of the bulk material; (b) diffraction is from nuclei, not the electrons surrounding it, so the positional accuracy is much better (10–13 m); and (c) all elements are detected with roughly equally sensitivity, since there is no Z^2 dependence for scattering of neutrons. The major uses of Neutron Diffraction are for structural refinement of atomic positions, particularly where light atoms are present with heavy atoms (e.g., O atom positions in superconductors), residual stress measurement (from diffraction line broadening), and determination of magnetic ordering. The major disadvantage of the technique is that it must be performed at a large government facility.

Physical adsorption isotherms involve measuring the volume of an inert gas adsorbed on a material's surface as a function of pressure at a constant temperature (an isotherm). Using nitrogen as the inert gas, at a temperature close to its boiling point (near 77K), such isotherms are used to determine the amount of the inert gas needed to form a physisorbed monolayer on a chemically unreactive surface, through use of the Brunauer, Emmett, and Teller equation (BET). If the area occupied by each physisorbed N_2 molecule is known ($16.2A^2$), the surface area can then be determined. For reactive clean metals, the area can be determined using chemisorption of H_2 at room temperature. Most clean metals adsorb one H atom per surface metal atom at room temperature (except Pd, which forms a bulk hydride), so if the volume of H_2 required for chemisorption is measured, the surface area of the metal can be determined if the atomic spacings for the metal is known. The main use of physical adsorption surface area measurement is to determine the surface areas of finely divided solids, such as oxide catalyst supports or carbon black. The main use of chemisorption surface area measurement is to determine the particle sizes of metal powders and supported metals in catalysts.

Raman Spectroscopy* 18

In Raman spectroscopy, a light beam of fixed wavelength (usually in the visible range) from a laser undergoes inelastic scattering as it interacts with the sample material. The photons can suffer energy loss if they excite vibrations in the sample. The energy losses, and therefore the vibrational frequencies involved, are determined by measuring the wavelengths of the scattered light. The value of an excited frequency is sensitive to the bonding between the atoms involved in that particular vibration.

Raman spectroscopy is closely related to infrared spectroscopy, which also determines the values of vibrational frequencies (see the FTIR summary), but it differs in the physics in that it involves a photon scattering process with energy loss, where as IR involves varying the photon energy and observing absorption occurring at the vibrational frequencies. Raman spectroscopy is better suited to optical microscopes, because of the single laser wavelength involved, and spatial resolution down to the one μm range is possible (cf. approximately a 10 μm limit for IR).

The depth probed depends very strongly on the optical properties of the material involved and the laser wavelength used. Moving into the UV range greatly reduces the probing depths, and allows a depth profiling capabilities if several wavelengths are used.

The major area of application for solids and liquids is chemical fingerprinting and the identification of unknown compounds. For solids, Raman is also used for phase identification, following amorphous/crystalline transitions, measurement of stress and strain, and, in the microscope mode, the detection and analysis of defects, including particles during wafer processing.

In Scanning Electron Microscopy (SEM), a probe electron beam (typically a few hundred eV to a few 10's keV energy) is finely focused (down to 10 Å capability in some instruments) and scanned over a solid surface. The interaction of the beam with the sample material generates a variety of responses, including fluorescence emission, which can be used for elemental analysis (see the EDS summary).

One of the major responses is a copious emission of low energy secondary electrons (0 to 50 eV range), which escape from the top few 10's Å of the material. The secondary electron yield strongly depends on the angle of impact between the probe beam and the local surface topography, so rastering the beam across the surface produces a changing intensity with changing topography. These changes in secondary electron emission intensity are used to modulate the brightness of a synchronously rastered cathode ray tube, creating an image. The image can be highly magnified up to 500KX, with a lateral resolution determined by the diameter of the probe beam, and have the look of an optical image (though the depth of focus is much greater).

In addition to the secondary electrons, a much smaller quantity of backscattered primary electrons are produced. Their intensity depends strongly on material (high for high Z elements), so the image also contains some Z contrast. Filtering out the low energy secondaries enhances this contrast (backscatter mode), producing an average Z dependant image instead of topography image.

The SEM is often the first or second (after an optical microscope) technique used to provide a magnified image of an area of the sample to be examined. It is often used in conjunction with an ancillary analytical technique, such as EDS, to provide elemental analysis capability to go with the imaging.

Scanning Transmission Microscopy (STEM) is a specialized form of TEM (see TEM summary) where the high energy electron beam (100 to 300 keV) is focused to a small spot (down to a minimum of about 3 Å) and scanned across a sample which is usually a material which has been sectioned to a very thin film (5 to 500 nm thick). After interaction with the material of the film, the transmitted beam can be subjected to various treatments, as in TEM (bright and dark field imaging; convergent beam electron diffraction, CBED; and electron energy loss spectroscopy, EELS). Also, the emitted X-ray fluorescence can be analyzed by EDS (see EDS summary). Keeping the sample to a very thin section stops the lateral spread of the primary beam as it passes through, allowing the spatial resolution to be determined by the diameter of the focus beam, down to almost atomic dimensions.

The major use of STEM is for imaging and compositional analysis (sometimes also chemical state analysis, from EELS) at very high spatial resolution across thin film material interfaces, such as found in semi-conductor and other high technology devices. Single atom imaging is possible for heavy atoms. The disadvantage is that a very thin interface cross-section must be first prepared. Also, at very high resolution, the electron beam is focused into such a small area that the local electron density becomes extremely high and can anneal, damage, or even burn through the film if the dwell time is too long.

In Scanning Tunneling Microscopy (STM), a sharp tip is brought to within a few Å of a conducting surface in ultra high vacuum and held at a positive or negative potential with respect to the tip. At this distance a quantum mechanical electron tunneling current flows between an individual atom on the surface and an atom on the tip (or vice versa, depending on the polarity). The magnitude of the current is an exponential function between the two atoms involved. So scanning the tip across the surface therefore produces a strongly varying current as the topography and distance between tip and surface atoms changes. This allows mapping of the topography (actually the surface electron density) with a resolution of 0.01 Å in the Z direction, and atomic resolution in the XY direction.

In the related technique, Scanning Force Microscopy (SFM; also known as atomic force microscopy, AFM), the tip is mounted on a cantilever and the Van De Waals forces between the surface and the tip deflects the cantilever. The deflection is also a strong function of the separating distance, allowing topography mapping, as in STM. The resolution in SFM is about ten fold poorer than STM, but the technique has the great practical advantages of being operated in air and not requiring a conducting sample.

The major use of STM is in the research area of imaging atomic structures on clean conducting and semi-conducting surfaces, under UHV conditions. SFM, on the other hand, is widely used throughout industry as a very high resolution surface profilometer to monitor surface roughness, defects, and man-made micro or nano structures.

In Solid State Nuclear Magnetic Resonance (NMR), a solid specimen is subjected to a strong magnetic field (typically 1 to 14 Tesla). If the nuclei present have unpaired spins, the magnetic field interacts with the magnetic dipole (spin), splitting the otherwise degenerate spin states into an upper (excited) state and a lower (ground) state. The energy of the splitting, for a given magnetic field, is characteristic of the nuclei concerned and is in the radio frequency range. If the sample is subjected to a pulsed radio frequency band covering this energy splitting, excitation from ground to excited state will occur. As the nucleus relaxes back to the ground state, it re-emits a radio signal at the frequency of the splitting (the NMR signal). Detection of the NMR frequency identifies the presence of the nucleus, and the intensity is proportional to the concentration. Small variations and splittings of the NMR frequency are related to bonding interactions with neighboring nuclei. While hydrogen nuclei NMR is by far the best known area of study and use (organic chemistry and MRI), about half the elements in the periodic table have unpaired nuclear spins suitable for NMR. The major use of solid state NMR is for element-selective phase identification and quantification, and for structural characterization of disordered phases which cannot be done by diffraction methods. A disadvantage of the technique is the very strong magnetic fields required, and that several different magnets might be required for different nuclei to get the NMR frequencies in the working range of the radio electronics. For this reason laboratories tend to concentrate on studies involving one particular element.

Static Secondary Ion Mass Spectrometry (Static SIMS)* 23

In Static Secondary Ion Mass Spectrometry (Static SIMS), a primary beam of ions (typically 1 to 10 keV) penetrates into a solid surface. The energy transferred to the lattice causes atoms, or clusters of atoms, to eventually be ejected from the surface region (from the top 1 to 3 atomic layers) as either ions or neutrals. These are mass analyzed in a mass spectrometer (either quadrupole or time-of-flight, TOF), allowing their identification. The difference, compared to dynamic SIMS (see the Dynamic SIMS summary), is that the dose of incidence primary ions is kept low (less than 5 times 1012 atoms/cm^2) so that, during analysis, the chemical integrity of the surface is maintained. Essentially every incoming ion strikes a fresh area not previously impacted. The objective is to deduce the chemistry of the outer layers by interpretation of the fragment ion pattern observed (cf. classical organic mass spectrometry).

The mass range accessible is up to 1000 amu with a quadrupole and 10,000 amu with TOF. TOF also has high enough mass resolution so that there is no ambiguity in atomic, or cluster identification (because atom isotopes have fractional masses, allowing any combination of masses to be distinguished with high enough mass resolution). The spatial resolution is controlled by the focus and energy of the ion beam, and is typically a few tens of µm, but sub-µm can be achieved in some cases.

The major use is in identification of the molecular species present at surfaces, particularly for polymers and organics. This may be for components of bulk materials (e.g., in polymer blends), or for monitoring changes caused by surface treatment, or for detection of contamination. Trace levels can sometimes be reached, but quantification is difficult and requires standards. With the high sensitivity and resolution of the TOF, the technique can also be used for detecting and quantifying trace heavy metals at surfaces.

Temperature programmed techniques are a series of catalyst characterization methods that involve thermal transient analysis. During temperature programmed experiments, a sample is exposed to different gaseous environments, and the sample temperature is increased linearly with time. The response of the system to the thermal transient is monitored by measuring: the sample weight change by ther-mogravimetric analysis (TGA); the change in temperature for constant heating by differential thermal analysis (DTA); the heat of reaction at constant temperature by differential scanning calorimetry (DSC); the consumed gas concentration by temperature programmed reduction or oxidation (TPR/TPO); or the evolved gas concentration by temperature programmed desorption or surface reaction (TPD/TPSR). The temperature programmed techniques provide information about the nature of solid compounds (TGA, DTA, and DSC), heat of reaction (DSC), reduction kinetics of metal oxides (TPR), oxidation kinetics of partially reduced metal oxides (TPO), adsorption/desorption kinetics (TPD), specific surface area of metallic catalysts (TPD), nature and number of surface acid sites (TPD), and reaction kinetics and mechanisms of surface intermediates (TPSR).

Range of elements measured	All solid elements
Sample requirements	Thermally stable solids
Sample form	All solid forms
Destructive	Only for thermally unstable solids
Chemical bonding information	Nature of surface intermediates and their bond strengths
Depth examined	Surface only (TPSR, TPD); bulk and surface (TPR, TPO, DSC, TGA, DTA)
Surface area limit	1 cm^2 for single crystals; above 1 m^2/gm for powders
Quantification	Can be calibrated
Main uses	Nature of surface sites and surface intermediates (reactivity and kinetics); nature of solid compounds
Instrument cost	Homemade or up to $100,000

In Transmission Electron Microscopy (TEM), a very high energy monoenergetic electron beam (100 to 400 keV) passes through a thin specimen (less than 1000 nm) of diameter less than 3 mm (necessary to fit in the electron optics column). A series of post specimen lenses transmits the emerging electrons, with spatial magnification up to 1,000,000, to a detector (fluorescent screen or video camera) viewed in real time.

Any regions of the sample under the incident beam (usually a few µm diameter) exhibiting crystallinity, will diffract electrons away from the central spot, forming a diffraction pattern observable at the back focal plane of the objective lens. Like X-ray diffraction, this can provide identification of crystalline phase, orientation, and lattice parameters. In micro-diffraction, the incident beam is focused down to sub-micron areas, but this focusing degrades the diffraction pattern.

The above mode of operation is termed *Diffraction Mode*. Another mode is *Imaging Mode*. In this mode the reconstructed, highly magnified image (up to 1,000,000) is observed at the image plane of the objective lens. By placing an aperture in the diffraction plane, either the electrons in the undeflected beam (the center spot) can be imaged (Bright Field Mode), or those in a deflected beam (the diffracted or scattered electrons) can be imaged (Dark Field Mode). An amorphous sample of uniform thickness can undergo only Z contrast scattering (higher Z, more scattering), so regions of higher Z show up as dark areas on a bright background in Bright Field Mode. Any crystalline region will also show up as a dark area. In dark field, regions of crystallinity, or higher Z, will show up as bright spots on a dark background.

Yet another mode of use is High Resolution TEM (HRTEM), in which atom positions can be established by collecting electrons from both the undeflected and diffracted beams and comparing the observed phase interference patters to a simulation.

TEM, with its many modes, and often involving ancillary materials analysis capabilities such as EDS (see EDS summary), is the mainstay of material science and analysis of small volume (areas and thickness). A fully equipped TEM laboratory will have several microscopes with differing capabilities, plus all the necessary sample preparation techniques. See also Scanning TEM (STEM), where the incident beam is focused down to almost atomic dimensions and scanned across the sample.

If monoenergetic photons in the ultraviolet and vacuum ultraviolet energy range strike a solid material, photoemission occurs by ejection of electrons from energy levels lower than that of the incident radiation. The kinetic energy of the ejected electrons is approximately given by the difference between the photon energy used and the Binding Energy of the electron level concerned. This is the Photoelectric Effect for which Einstein was awarded a Nobel Prize. Ultraviolet Photoelectron Spectroscopy (UPS) is the measurement of the kinetic energy distribution of the ejected electrons. The physics of UPS is the same as XPS, except that because much lower photon energies are used (usually the He resonance lines at 21.2 eV and 40.8 eV in the laboratory), the electron energy levels accessed are mostly restricted to the valence levels. The valence level energy distributions are characteristic of the bonding in the whole material but generally do not provide an atomic identification, instead giving detailed information about the electronic structure. Some atoms have low lying core levels, characteristic of the atom concerned, and if these fall within the accessible range of the photon energy used, then atomic identification and quantification from intensities is possible. If synchrotron radiation is used as the photon source, the "ultraviolet" photon energy range used can be extended up to above 100 eV, making many more low lying core levels accessible. Both the laboratory and synchrotron photon sources result in the kinetic energies of the photoelectrons being much lower than that typically in XPS, which results in greater surface sensitivity. Using synchrotron radiation, the sensitivity can also be tuned by varying the photon energy. Spectral resolution in UPS is higher than in XPS, because the line width of the photon source is much less. The main uses of UPS are for detailed electronic structure study of well-defined solid surfaces (band structure), and for adsorption on solid surfaces, and molecular structure information on polymers and inorganic compounds. In cases where increased surface sensitivity and spectral resolution are useful for materials analysis, compared to XPS, UPS access to available core levels is a definite advantage. Laboratory UPS photon sources are easily added to XPS instrumentation.

In X-Ray Diffraction (XRD), a collimated beam of mono-chromatic X rays between 0.5 Å and 2 Å wavelength strike a sample and are diffracted by crystal planes present. Bragg's law,

$$\lambda = 2d \sin\theta$$

relates the spacing between planes, d, to the diffraction angle, 2θ, which is scanned to pick up diffraction from the different crystal planes present. The azimuthal orientation of the different beams also reveals the crystalline orientation. If the material is poly-crystalline, then diffraction rings, instead of a spot pattern, are formed (powder diffraction). Distortions or broadenings of the diffraction beams carry information on crystal strain and grain size.

Because X rays penetrate deeply (strongly Z dependent; at $\lambda = 1.54$ Å the absorption length is 1 mm for carbon, 66 μm for Si, and 4 μm for iron), XRD is intrinsically a bulk technique. Typically, however, large areas are used (several mm diameter), and there is sufficient intensity and spectral resolution using modern standard equipment to study films of thickness down to the 100's nm range for high Z material. If specialized instrumentation and geometries are used (Grazing Angle XRD, Double Crystal Diffraction XRD, synchrotron radiation source), sensitivities down to 10's nm are easily achieved, and mono-layer sensitivity is possible. Small area capabilities also exist (micro-beam XRD), but then thicker samples are required to compensate for intensity loss of the smaller areas.

The major uses of XRD are identification of crystalline phases, determination of strain, crystalline orientation and size, epitaxial relationship, and the accurate determination of atomic positions (better then in electron diffraction). Because of the strong Z dependence of X-ray scattering, light elements are difficult to deal with, particularly in the presence of heavy elements.

In X-Ray Fluorescence (XRF), the mono-energetic X-ray beam from a vacuum X-ray tube (W, Mo, Cr, and other targets are used to provide a range of energies) irradiates the sample, in air, causing excitations of core electrons of the atoms present. As these excited atoms decay back to their electronic ground states, light is emitted in the X-ray region. The specific wavelengths emitted are characteristic of the energy levels involved, and therefore of the atoms concerned. The X-ray fluorescent wavelengths are determined using a crystal diffractometer.

The depth probed is Z dependent, but, since X rays are involved both in the excitation and the emission, it is many μm, making XRF a bulk technique. A specialized grazing incidence (total reflection) adaptation exists, however, for analyzing material at the surfaces of semi-conductor wafers (see TXRF summary). Also, there is sufficient total intensity such that thin films (μm level or lower, depending on instrumentation and Z) can be analyzed provided there are no interfering elements in the substrate. Standard instruments have no significant spatial resolution (few mm), but micro-beam systems down to 10 μm do exist. The technique is applicable to all elements except H, He and Li. Energy dispersive X-ray analysis (see EDS summary) is a closely related technique.

The major uses of classical XRF, in air, are the identification of elements and the determination of composition for bulk materials. For thin films intensity/composition/thickness equations are used to determine the composition and thickness of individual layers in single or multi-layered stacks, such as used in the disk drive and semi-conductor industry.

X-Ray Photoelectron Spectroscopy (XPS) provides atomic and chemical state identification from the measurement of the BE's of the atomic core level electrons present in a sample, which are characteristic of the atom concerned (see XPS). If the sample is single crystal, the angular distribution of the ejected photoelectron also carries structural information (also true for Auger electrons), which originates from the interference effects as the outgoing photoelectron is scattered by neighboring atoms in the material. The electron–atom scattering process focuses electrons in the forward direction along atom–atom directions, leading to the simple observation that intensity is peaked in directions corresponding to rows of atoms. An angle resolved electron energy analyzer is used to map the intensity distribution as a function of angle in an element specific manner. By modeling the results it is possible to determine bond angles to within 1 degree and bond lengths to within 0.05 Å. The technique is surface sensitive, since it is an XPS or AES measurement (5 to 50 Å). In the AED mode, the use of the structural information can have lateral resolution of as good as 150 Å (like AES in general). The main use of the technique is to determine adsorption sites and thin film epitaxial growth relationships in an element specific manner.

In X-Ray Photoelectron Spectroscopy (XPS), also known as Electron Spectroscopy for Chemical Analysis (ESCA), mono-energetic soft X rays (usually Al K alpha at 1489 eV or Mg K alpha at 1256 eV) bombard a solid sample in ultra high vacuum. One of the interaction processes occurring is the ejection of photoelectrons, according to the Einstein Photoelectric Law:

$$KE = h\nu - BE$$

where $h\nu$ is the incident X-ray energy, KE is the kinetic energy of the photoelectron, and BE is the binding energy of the electron in the particular energy level of the atom concerned. Measurements of the KE's of the ejected photoelectrons directly determines the BE's, which, for core levels of atomic orbitals, are characteristic of the atoms concerned and therefore identify their presence. On a finer energy scale, small "chemical shifts" in the BE's provide chemical state identification (such as the oxidation state for a metal). The technique is applicable to all elements, except H and He, since they do not have characteristic atomic core levels.

For a solid, XPS probes 2 to 20 atomic layers deep, depending of the KE of the ejected electron, the angle w.r.t., the solid surface of detection, and the material. XPS is therefore a true surface technique. If measurement is made as a function of angle, depth distribution information is available over the depth probed. XPS is also used with sputter depth profiling to go beyond these depths. Modern laboratory XPS instruments can provide *practical* lateral resolution capability down to the 10 to 20 µm range. Specialized systems can go lower, but acquisition time becomes very long.

The particular strengths of XPS are semi-quantitative elemental analysis at surfaces without standards, quantitative analysis with standards, and chemical state determination for materials as diverse as biological to metallurgical.

Index

particle size,   94–95
preparation,   90–92
promotion effects,   101–102
stable phases,   92–94
sulfur vacancies,   104–105
surface area,   94–95, 96, 102
surface composition,   97–100

Catalysts   *See* alumina pillared clays, bulk metal oxides, bulk metals, bulk metal sulfides, molecular sieves, supported metal oxides, supported metals, and supported metal sulfides.
characterization of catalysts
    atomic distances,   32
    binding energy,   38–42
    bonding of ferromagnetic metals,   43
    bulk chemical analysis,   4–5
    bulk composition,   57, 139–141
    bulk structure,   51–58
    chemical species,   31
    contact potential,   12
    coordination numbers,   12, 94
    coverage of support,   73–75
    crystallites,   71–72, 113
    dispersion,   20, 22
    gas-surface interactions,   11–12
    interatomic distances,   12
    microstructure,   6–8, 58, 61
    molecular information,   12
    morphology,   6–8, 61, 77, 94–95
    oxidation states,   31, 41, 57–60, 63–64, 75–77
    particle size,   27–28, 29, 42, 94–95
    pore size and volume,   8–9, 62, 137–138, 143–144, 155–158
    stacking of layers,   153
    surface acidity,   158
    surface area,   8–9, 19–20, 61–62, 94–95, 96, 119–120, 155–158
    surface chemistry,   77–79
    surface composition,   9–11, 63–64, 97–100
    surface structure,   12–15
    void volume,   142
chemical analysis,   4–5
chemisorption   *See also* temperature-programmed desorption.
    for bulk metal oxides,   64
    for bulk metals,   9

for bulk metal sulfides,   96, 102–103
flow method,   9
pulse method,   9, 23, 64
static volumetric,   9, 21–22
stoichiometry,   9, 20
for supported metal oxides,   78
for supported metals,   19–25
for supported metal sulfides,   120
clays   *See* alumina pillared clays.
cluster size   *See* particle size.
cobalt in sulfidic catalysts,   116–119
comaceration,   91
computer simulations,   7–8
contact potential,   12
coordination numbers,   12, 94
coprecipitation,   50
coverage of support,   73–75
crystallites,   71–72, 113

Dark-field   *See* annular dark-field imaging.
diffuse reflectance spectroscopy (DRS),   12
dispersion,   20, 22
DOR NMR   *See* double rotation NMR.
double rotation NMR (DOR NMR),   141–142
DRS   *See* diffuse reflectance spectroscopy.

EDX   *See* energy dispersive X-ray analysis.
EELS   *See* electron energy loss spectroscopy.
electron energy loss spectroscopy (EELS),   6
electron paramagnetic resonance (EPR),   76
electron probe microscopy (EMPA),   5
electron spin resonance (ESR),   58, 95–96
EMPA   *See* electron probe microscopy.
energy dispersive X-ray analysis (EDX),   6
EPR   *See* electron paramagnetic resonance.
ESR   *See* electron spin resonance.
EXAFS   *See* extended X-ray absorption fine-structure spectroscopy.
extended X-ray absorption fine-structure spectroscopy (EXAFS)
    for bulk metal oxides,   59–60
    for bulk metals,   12
    for bulk metal sulfides,   94, 96, 99
    for supported metal oxides,   72, 76–77, 81, 82
    for supported metals,   32–36
    for supported metal sulfides,   115–116, 119, 122, 123

UPS   *See* ultraviolet photoelectron
        spectroscopy.
UV photoemission,   40
UV source,   40

Video recorders,   8
void volume,   142

Work function,   12

XANES   *See* X-ray absorption near-edge
        structure spectroscopy.
XAS   *See* X-ray absorption spectroscopy.
XPD   *See* X-ray powder diffraction.
XPS   *See* X-ray photoelectron spectroscopy.
X-ray absorption near-edge structure
        spectroscopy (XANES)
    for bulk metal oxides,   59–60
    for bulk metal sulfides,   96
    for supported metal oxides,   75–77,
        81, 82
    for supported metals,   31
X-ray absorption spectroscopy   *See* extended
        X-ray absorption fine-structure.
        spectroscopy, near-edge X-ray absorption
        fine-structure spectroscopy, and

X-ray absorption near-edge structure
        spectroscopy.
X-ray diffraction (XRD)
    for alumina pillared clays,   152–153, 155
    for bulk metal oxides,   51–52, 65
    for bulk metal sulfides,   93, 94
    line-broadening in,   5–6, 27–28, 153
    for molecular sieves,   137–138
    for supported metal oxides,   71
X-ray fluorescence (XRF)
    for bulk metal oxides,   57
    for bulk metals,   5
    for molecular sieves,   139
X-ray maps,   6
X-ray photoelectron spectroscopy (XPS)
    for bulk metal oxides,   63–64, 65
    for bulk metals,   10
    for bulk metal sulfides,   96, 97
    for supported metal oxides,   73–74, 75
    for supported metals,   38–40
    for supported metal sulfides,   120
X-ray powder diffraction (XPD),   133–135
XRD   *See* X-ray diffraction.
XRF   *See* X-ray fluorescence.

Zeolite identification,   134–135
zeolites   *See* molecular sieves.